THE NUCLEAR WEAPONS COMPLEX

Management for Health, Safety, and the Environment

Committee to Provide Interim Oversight of the
DOE Nuclear Weapons Complex
Commission on Physical Sciences, Mathematics, and Resources
National Research Council

National Academy Press
Washington, D.C. 1989

NOTICE: The project that is the subject of this report was approved by the Governing Board of the National Research Council, whose members are drawn from the councils of the National Academy of Sciences, the National Academy of Engineering, and the Institute of Medicine. The members of the committee responsible for the report were chosen for their special competences and with regard for appropriate balance.

This report has been reviewed by a group other than the authors according to procedures approved by a Report Review Committee consisting of members of the National Academy of Sciences, the National Academy of Engineering, and the Institute of Medicine.

The National Academy of Sciences is a private, nonprofit, self-perpetuating society of distinguished scholars engaged in scientific and engineering research, dedicated to the furtherance of science and technology and to their use for the general welfare. Upon the authority of the charter granted to it by the Congress in 1863, the Academy has a mandate that requires it to advise the federal government on scientific and technical matters. Dr. Frank Press is president of the National Academy of Sciences.

The National Academy of Engineering was established in 1964, under the charter of the National Academy of Sciences, as a parallel organization of outstanding engineers. It is autonomous in its administration and in the selection of its members, sharing with the National Academy of Sciences the responsibility for advising the federal government. The National Academy of Engineering also sponsors engineering programs aimed at meeting national needs, encourages education and research, and recognizes the superior achievements of engineers. Dr. Robert M. White is president of the National Academy of Engineering.

The Institute of Medicine was established in 1970 by the National Academy of Sciences to secure the services of eminent members of appropriate professions in the examination of policy matters pertaining to the health of the public. The Institute acts under the responsibility given to the National Academy of Sciences by its congressional charter to be an adviser to the federal government and, upon its own initiative, to identify issues of medical care, research, and education. Dr. Samuel O. Thier is president of the Institute of Medicine.

The National Research Council was organized by the National Academy of Sciences in 1916 to associate the broad community of science and technology with the Academy's purposes of furthering knowledge and advising the federal government. Functioning in accordance with general policies determined by the Academy, the Council has become the principal operating agency of both the National Academy of Sciences and the National Academy of Engineering in providing services to the government, the public, and the scientific and engineering communities. The Council is administered jointly by both Academies and the Institute of Medicine. Dr. Frank Press and Dr. Robert M. White are chairman and vice chairman, respectively, of the National Research Council.

Library of Congress Catalog Card No. 89-63691
International Standard Book Number 0-309-04179-1

Additional copies of this report are available from:

National Academy Press
2101 Constitution Avenue, NW
Washington, D.C. 20418

S068

Printed in the United States of America.

First Printing, December 1989
Second Printing, January 1990
Third Printing, May 1990
Fourth Printing, December 1990

Committee to Provide Interim Oversight of the DOE Nuclear Weapons Complex

RICHARD A. MESERVE, Covington & Burling, Washington, D.C., *Chairman*
RONALD L. ATKINS, Naval Weapons Center
ALBERT CARNESALE, John F. Kennedy School of Government
JESS M. CLEVELAND, U.S. Geological Survey
DAVID G. HOEL, National Institute of Environmental Health Services
GEORGE M. HORNBERGER, University of Virginia
PAUL KOTIN, Denver, Colorado
DENNIS J. KUBICKI, U.S. Nuclear Regulatory Commission
J. CARSON MARK, Los Alamos, New Mexico
MICHAEL R. OVERCASH, North Carolina State University
WOLFGANG K.H. PANOFSKY, Stanford Linear Accelerator Center
RICHARD L. SAGER, JR., Lithium Corporation of America
RICHARD B. SETLOW, Brookhaven National Laboratory
DAVID R. SMITH, Santa Fe, New Mexico
STEWART W. SMITH, Kirkland, Washington
JOHN E. TILL, Radiological Assessments Corporation
VICTORIA J. TSCHINKEL, Landers, Parsons, & Uhlfelder, Tallahassee, Florida
F. WARD WHICKER, Colorado State University
WILLIAM L. WHITTAKER, Carnegie-Mellon University
GEROLD YONAS, Titan Technologies (resigned May 1989)

Staff
STEVEN M. BLUSH, Project Director (until February 1989)
MYRON F. UMAN, Project Director (beginning February 1989)
SARAH CONNICK, Program Staff Officer
PATRICK D. RAPP, Program Staff Officer
JAROSLAVA P. KUSHNIR, Administrative Specialist
ALFREDA B. McELWAINE, Administrative Secretary
CAROLYN M. BATTLE, Senior Secretary

Commission on Physical Sciences, Mathematics, and Resources

Preface

In the aftermath of the accident at the Chernobyl nuclear power station, the Department of Energy (DOE) asked the National Research Council to examine possible implications of the accident for the large reactors operated by the Department. The reactors included those then operating at the Savannah River Site in South Carolina and at the Hanford Nuclear Reservation in the state of Washington that were used in the production of special materials for nuclear weapons, as well as those used in testing and research. In response, the National Research Council issued reports (1987 and 1988b, respectively) that focused on a variety of safety, management, and technical issues.

In the meantime, concerns developed with regard to the other, nonreactor facilities in the nuclear weapons complex. These facilities include 17 installations throughout the United States that are engaged in the range of activities required to produce nuclear weapons: designing them; processing materials for uranium enrichment; preparing materials for irradiation in the production reactors; processing materials from the reactors; producing components for weapons; assembling the components into a completed device; and testing components and devices. As a result of the concerns, Congress directed the Secretary of Energy to request that the National Research Council report its conclusions and recommendations concerning health, safety, and environmental issues arising throughout the complex and steps that would enhance the safety of operations at the facilities (see Appendix F).

This report fulfills the Secretary's request. It was prepared by a committee, appointed for the purpose by the National Research Council, whose members brought to their task expertise across a spectrum of pertinent disciplines in health, safety, and environmental matters (see Appendix A).

In conducting its study, which began in August 1988 and extended through September 1989, the committee reviewed extensive documentation provided by the Department and its contractors and engaged in briefings and discussions with a variety of others who are knowledgeable about the complex, including personnel from the U.S. Environmental Protection Agency and state agencies. The committee also made site visits to several of the facilities, including the Hanford Nuclear Reservation in Washington; the Y-12 Plant in Oak Ridge, Tennessee; the Rocky Flats Plant in Colorado; the Los Alamos National Laboratory in New Mexico; the Lawrence Livermore National Laboratory in California; the Idaho National Engineering Laboratory; the Savannah River Site in South Carolina; and the Pantex Plant in Texas.

During the course of the study, a number of developments affected our work. First, the operation of the weapons complex came under increasingly intense public scrutiny and criticism. News articles concerning the complex appeared almost daily, several congressional hearings were held, and a wealth of detailed commentary about the complex was offered by a variety of individuals and organizations, such as the General Accounting Office and the Advisory Committee on Nuclear Facility Safety. These reports not only served as an important source of information, but also reinforced the need for the committee to step back and view the complex in broad overview.

Second, the national administration and the upper management of DOE changed in early 1989. The Secretary of Energy, James D. Watkins, publicly expressed his dismay at the past performance of the Department in managing the weapons complex and stated his intention to make substantial, if not radical, changes. He has already introduced some changes and has indicated that further change will be forthcoming. He has also acknowledged the extent of local, state, and federal jurisdiction in matters related to health, safety, and the environment. We welcome the dynamic new leadership of the Department, of course, but the fact that the complex was changing as we studied it served to complicate our task. We have sought to prepare a timely report that accurately reflects current circumstances, but we recognize the possibility that in some cases events may have overtaken us even as we were completing our work, and we cannot yet fully assess the significance of those changes of which we were aware.

In spite of the wealth of experience we brought to our task, we faced a number of limitations. We could not and did not examine the entirety of the nuclear weapons complex. Some elements of the complex—the production reactors, the Nevada test site, the gaseous diffusion plants, the assembly areas at Pantex, and nuclear waste facilities—were excluded from our purview by the Department, as was transportation of materials between sites. In light of our schedule, we agreed that our examination would be more useful if we were to focus attention on the principal remaining facilities. Although we believe the report provides an appropriate and useful overview of the complex, it is not a complete study of the complex, of any individual facility, or even of any specific building within a facility. Such studies would be immense undertakings, and no committee serving pro bono on a part-time basis could hope to accomplish them.

Although the scope of our examination of the complex was necessarily limited, we nevertheless believe that our findings are broadly applicable. In no way, however, does our report pretend to provide a complete inventory of the health, safety, and environmental issues facing the DOE nuclear weapons complex. It remains for the Department and others to build on our work in what must be viewed as a continuing effort.

Our report is also framed by the expertise and knowledge of the committee members. The diversity of the facilities under study required a committee comprising individuals with disparate technical backgrounds. For the study to be kept manageable, however, there was a practical limit on the areas of expertise that could be represented.

In addition, we relied strongly on data provided by the Department and its contractors. They were responsive to our requests for information, but our firsthand data gathering had to be limited to what could be gleaned from our brief site visits and our meetings with contractor and DOE staffs.

For these reasons the term "oversight" committee is perhaps a misnomer for our role. We were not asked for, nor did we give, continuing advice, let alone direction, to the Department, its contractors, or any of their personnel. This report is our only output.

We have not examined the financial costs that would be incurred in remedying existing deficiencies in health and safety measures, in bringing environmental protection up to applicable standards, or in redressing environmental problems created by past activities. Many estimates of this sort have been disseminated by the media. We do not believe, however, that such estimates are meaningful without the formulation of specific plans and policies, and we neither endorse nor contradict any published figures. Nevertheless, it is clear that substantial funds will have to be spent to accomplish the objectives.

We have not attempted to reach any bottom-line conclusion as to whether the operations are "adequately safe." Such a judgment would require a level of scrutiny of operations beyond the capacities of a committee like ours. Moreover, acceptable risk must ultimately be measured by balancing the benefits of the activities against their costs. Here the "benefit" is the supply of special nuclear materials and nuclear weapons; the "cost" is measured in both financial terms and in less quantifiable health and environmental terms. Evaluation of the balance requires societal judgments of a sort we were not asked to make and for which, even if asked, we could not have claimed any special expertise.

The committee also did not examine the basis of national security requirements that translate into the demand for the materials produced by the facilities. There was much public discussion during the period of our deliberations about the need for continued production of plutonium and the immediacy of the need for facilities to produce tritium. These matters are important and require prompt resolution, but examination of the demand for materials raises matters of national security policy that, again, extend beyond our charge.

We are aware of the disclaimers that flavor this preface, but they are caveats that must be understood in an undertaking like this. The immensity of our task caused us to approach it with trepidation, but we end it with a sense of satisfaction and with the hope that our efforts will prove helpful to the Department and the Congress. It only remains to be said that we could scarcely have completed the work without help from many sources. The National Research Council staff assigned to the project gave of their time and talents with energy and enthusiasm, and we are in their debt. We are also grateful to the Department of Energy and its contractors and to the many others too numerous to name here who assisted us in our work.

Richard A. Meserve, *Chairman*
October 1989

Contents

develop innovative technologies for waste handling. In particular, DOE's consideration of waste management should include all production wastes and residuals, not only those involving radioactivity. Waste minimization principles should be applied across the complex. Process modifications to minimize waste generation often require significant front-end capital investment, but they can be cost effective in the long term by reducing overall waste management expenses. An economic framework for evaluating such issues is needed.

SAFETY

Viewed from the perspective of conventional U.S. industry, the weapons complex has an excellent occupational safety record. Workdays lost because of injury per worker-hour are considerably fewer than those observed in the private sector as a whole. Nonetheless, there is room for improvement. Indeed, DOE and its contractors must be especially vigilant because of the unique hazards that arise from the materials that are handled in the complex and the special risks that could arise if a catastrophic event, such as a fire or earthquake, were to release hazardous materials.

We identified a variety of safety matters that deserve examination by DOE and its contractors. We cannot claim, however, that our scrutiny was so extensive or thorough that other similar hazards might not be found on more probing scrutiny. Clearly, some circumstances are more serious than others; we believe three particular areas deserve priority attention.

First, the organization, staffing, training, and equipment of the site fire departments are, with few exceptions, superior; but there are troublesome elements in the fire protection program. For example, although DOE fire protection criteria require that fire suppression equipment be installed in locations where a fire could cause damage to critical equipment and interrupt production for extended periods, we found instances in which such protections were not provided. The protection criteria should also cover safety systems.

Second, efforts to control the inhalation of radioactive materials, particularly at Rocky Flats, have led to an unwarranted reliance on respirators. Respirators are uncomfortable; increase fatigue; and impair employee alertness, efficiency, oral communication, and morale. Moreover, they are not effective over the long term in preventing radioactive inhalation problems. In our view, a pattern of routine use of respirators is an indication of the failure of production, maintenance, and housekeeping procedures.

Plutonium exists in the exhaust ducts downstream of the high-efficiency particulate air (HEPA) filters at the plutonium finishing facility at Hanford. Kilogram quantities of plutonium have also accumulated downstream of the HEPA prefilters in an exhaust duct of Building 771 at Rocky Flats. It is noted that similar problems may exist elsewhere in the complex.

Third, while strong nuclear safety programs exist throughout the complex, there are weaknesses in the program for controlling the more conventional industrial hazards.

HEALTH

The production of nuclear weapons involves activities and materials that can affect human health adversely. Indeed, some of the hazards present in the work environment, such as those arising from certain weapons materials, are unique. An occupational health program is aimed at preventing illness, diagnosing and treating illness, and monitoring the exposure of workers to hazardous chemicals and radioactive substances. In our view, programs in the complex for the occupational health of employees of DOE and its contractors need significant improvement.

Although medical attention is provided to employees with injuries or potentially harmful exposures to radiation or chemicals, medical departments are rarely involved in decisions related to monitoring and controls in the workplace. Medical personnel must often rely on their patients to identify the chemicals to which they have been exposed. In short, the medical departments are for the most part relegated to a reactive role. Although the central focus of programs in health physics, safety, industrial hygiene, and emergency planning is the protection of human health, medical input has been slight. DOE should assure that there is collaboration in all these matters involving medical expertise, and that the medical department in DOE headquarters is given sufficient resources to administer, monitor, and effect changes in these programs.

In addition to improving the protection programs, it is essential to monitor workers to assess the effectiveness of the programs, to identify opportunities for improvement, and to allow for timely medical intervention. Medical monitoring and surveillance programs in the complex should be improved substantially through the use of standardized protocols for data collection, storage, and analysis.

There is considerable concern in the general public about the risks arising from exposure to radiation. Although the best available estimates suggest that the radiation risks arising from past and current releases of radiation from the complex are low enough not to have resulted in any significant increase in risk to the public, more detailed assessment of the effects of these exposures is of both scientific and public interest. DOE should therefore continue to support epidemiologic studies to examine the facilities' effects, if any, on the public. To be credible, the studies should be designed and directed outside DOE and subject to external peer review. Moreover, DOE should compile data on its workers in a comprehensive data base and, with these data, continue to support peer-reviewed epidemiologic research.

MODERNIZATION

Much of the physical plant of the weapons complex is old, and many of the processes used in production are outdated. Opportunities—and challenges—exist not only in refurbishing the complex, but also in introducing alternative processes that could improve overall efficiency and facilitate the attainment of health, safety, and environmental, as well as production, goals. At the request of Congress, DOE prepared a report on the modernization of the complex that placed a high priority on the refurbishment of the plutonium recovery capacity in Building 371 at the Rocky Flats Plant and in the construction of a special isotope separation (SIS) facility for separating weapons-grade plutonium from spent reactor fuel at the Idaho National Engineering Laboratory. Although we were not in a position to assess current or future requirements for weapons material, it is apparent that the current supply of plutonium and the current capacity to process both virgin and recycled plutonium from retired weapons and scrap are adequate to meet the demand for plutonium for weapons of the number and general character currently in the national stockpile. Because plutonium is costly, long-lived, toxic, and must be carefully safeguarded, it is not sensible to produce more weapons-grade plutonium than is reasonably needed. The renovation of Building 371 is questionable in light of the plutonium recovery capacity that exists elsewhere in the complex. Not only may SIS be unneeded, it also presents important new considerations relating to safety and nuclear safeguards. The committee thus urges the Department to reconsider its plans relating to these facilities.

It is inevitable, however, that DOE will eventually be obliged to replace or renovate some of its aging facilities. In undertaking this modernization, clear opportunities exist to make more effective use of technology. They include the opportunity to eliminate or minimize fluoride-based plutonium processing and to make more effective use of advanced metal-forming, computer, communications, and robotics technologies. DOE and its contractors must be alert to opportunities to introduce new technology, to reduce the generation of wastes, or to employ less hazardous materials, thereby improving the effectiveness of the complex in meeting production goals in a way that is consistent with health, safety, and environmental goals.

1
Introduction

Nuclear weapons are central to the national security policy of the United States. Whatever our personal feelings about these weapons, we recognize their importance now and in the years ahead. The weapons exist, and we will be the custodians of them and the materials of which they are made indefinitely into the future.

The Department of Energy designs, manufactures, and maintains nuclear weapons for the Department of Defense. The weapons depend on the unique properties of isotopes of certain elements, among them uranium-235, plutonium-239, and tritium (the isotope of hydrogen with an atomic weight of 3). To produce a nuclear weapon, these materials must be configured so that at the appropriate instant they are brought together in a fashion that leads to the release of enormous amounts of energy in a very brief period of time (see Appendix E). Uranium-235 exists in nature, but it must be isolated from other uranium isotopes. Plutonium and tritium do not exist in nature in significant quantities, so they are "created" in production nuclear reactors. The Department creates, maintains, and modernizes the national stockpile of nuclear weapons at government-owned, contractor-operated facilities that, taken together, make up the U.S. nuclear weapons complex.

THE NUCLEAR WEAPONS COMPLEX

Some 17 major facilities in 12 states are engaged in the production of nuclear materials and their assembly into weapons. The total budget for the operation of these plants in FY 1990 is nearly $10 billion.

A description of the various facilities and their respective roles in the weapons complex is given in Appendix B. In brief overview, the facilities are of three different types: weapons laboratories, materials production facilities, and weapons production facilities (see Table 1.1). The weapons laboratories design and develop the weapons and test the various components and devices. The materials production facilities are engaged chiefly in the production of special nuclear materials. Much of the focus of activities in the complex is on the preparation of materials for transmutation in the production reactors and the subsequent extraction and purification of plutonium and tritium from reactor targets and recycled materials.

The weapons production facilities fabricate the required nuclear components, provide the various nonnuclear components, and assemble the weapons. The nonnuclear components include various electrical and mechanical devices, conventional explosives, neutron generators, shielding, and other parts.

The facilities in the complex are operated by contractors supervised by DOE. Most of the technical expertise with regard to the design and operations, as well as the detailed knowledge of the facilities, necessarily resides with the contractors. DOE is responsible for assuring that the demands for production are satisfied; that health, safety, and environmental concerns are adequately met; and that security and safeguard issues are appropriately addressed. The Department must also ensure that public funds are appropriately spent. In fulfilling these responsibilities, DOE and its contractors maintain a staff of about 80,000 people.

The weapons complex faces two types of hazards: those confronted by any large industrial complex, and the special hazards that arise from the unique

TABLE 1.1 Nonreactor Facilities in the Weapons Complex, Other Than Test Sites and Waste Repositories

Weapons Laboratories	Materials Production Facilities	Weapons Production Facilities
Los Alamos National Laboratory	Ashtabula Plant	Kansas City Plant
Lawrence Livermore National Laboratory	Feed Materials Production Center	Mound Facility
Sandia National Laboratory	Hanford Nuclear Reservation	Pantex Plant
	Idaho Chemical Processing Plant	Pinellas Plant
	Oak Ridge Gaseous Diffusion Plant	Rocky Flats Plant
	Paducah Gaseous Diffusion Plant	Y-12 Plant
	Portsmouth Gaseous Diffusion Plant	
	Savannah River Site	

mission of the weapons complex. The familiar hazards include those posed by fire, electrical and rotating machinery, compressed-gas systems, and the handling of high-energy explosives and hazardous chemicals, such as hydrogen fluoride, mercury, and various solvents used in processing. The unusual hazards derive from the radioactivity or chemical toxicity of some of the special materials required for weapons or incidental to their processing and handling and the unique and absolute need to avoid accumulations of plutonium or enriched uranium that could result in a criticality accident (see Chapter 4 and Appendix C). Thus fire is a conventional industrial hazard, but it can create special problems in facilities processing, for example, plutonium.

The use of these materials and the need to limit the exposure of humans to them put special demands on air supply, filtering, and monitoring. Care must be taken to assure that large quantities of these materials are not released to the environment in the event of an error or accident. Of course, there is also the need to control and monitor effluents from the plants and to assure that the disposal or storage of wastes does not have detrimental effects on people or the environment.

The assurance of satisfactory operations depends on many factors: proper design and choice of materials for construction and equipment; careful attention to proper procedures; awareness of hazards and how to avoid them; careful maintenance and the ability to upgrade aspects of a facility when the need arises; use of design, procedures, and training to avoid and mitigate accidents; and a thorough dedication to health, safety, and environmental compliance at all levels.

THE CURRENT SITUATION

The Department (and ultimately the Congress and the U.S. public) now confronts a serious challenge in managing and operating the weapons complex. Most of the facilities were built before the mid-1960s, and many are now approaching the end of their useful lives. Some of the facilities were constructed on a wartime crash basis. At that time, little consideration was given to design for severe earthquakes, maximum probable floods, tornado-borne objects, or other extreme conditions for which much of the data, as well as the techniques for taking such events into account, have been developed only in more recent decades. Many facilities have obsolete equipment, and the difficulties of continued reliance on such equipment have been compounded by inadequate attention to maintenance over long periods of time.

Current operations are also burdened by an environmental legacy derived from past operations. The primary focus of attention in the early years of operations—with respect to safety of operations as well as the handling of effluents and wastes—was on radioactive materials. The customary practices of the time were followed with regard to nonradioactive wastes. Now, not only has the behavior of the radioactive effluents and wastes proven to be far more troublesome than anticipated but also the handling of the more familiar nonradioactive effluents and wastes has been shown to be seriously deficient. As a result there are now potentially serious environmental problems throughout the complex, and substantial pressure has arisen to restore the environment—a prospect that is both technically and financially challenging.

The difficulties that the Department now confronts are exacerbated because the general climate within which the complex operates has recently changed in several significant respects. First, the Department must now operate under greater public scrutiny than it did in the past. The cloak of secrecy that shrouded many of its operations in the past is lifting. The revelations of environmental contamination from past activities and public concern with nuclear weapons and with nuclear activities in general have combined to create a climate of distrust. The intense public scrutiny is not likely to abate.

Second, DOE now has less control over its operations than it did in the past. The Department has acknowledged that it will comply with the standards established and enforced by other agencies of government (both state and federal), and it is no longer in a position to define its own environmental and safety obligations independently. The outside agencies have no direct concern for DOE's production goals, yet they must now be satisfied that the complex is adequately meeting its environmental obligations.

Third, the safety and environmental standards with which the facilities must comply have become increasingly stringent over time and may become even more stringent in the future. The trend presents particular difficulties in the operation of aging plants that were designed without consideration of such standards.

Fourth, the budgetary environment within which the Department must operate has changed to one of stringency and constraints, and competition for the use of funds appropriated for nuclear weapons production has become intense. Moreover, an increasing portion of the budget for weapons-related activities is likely to be used for environmental remediation. In fact, the Department in many instances may be obligated by legally enforceable orders to allocate discretionary funds to environmental cleanup.

Fifth, although nuclear weapons are expected to play a continuing role in deterring war, continued nuclear weapons production is perceived to have decreasing significance in the overall national security of the United States. Significant reductions in nuclear weapons are desired by many, and uncertainties exist about the status of the weapons complex as an element essential to U.S. national security.

THE CHALLENGES

In the remainder of this report, we seek to illuminate some of the issues that confront the nation regarding the nuclear weapons complex. In broad overview, our findings and recommendations bear upon the following overarching challenges.

Setting Production Goals

The "demand" for production arises from the requirements for nuclear weapons that are established in the Presidential Stockpile Memorandum. To generate the memorandum, DOD, in consultation with DOE, interprets fundamental national security objectives with regard to nuclear deterrence under broad White House guidance. Obviously, the demand for weapons provides the fundamental underpinning for the activities of the weapons complex. But, although DOD plays a dominant role in defining the requirements for nuclear weapons, it is DOE that is obligated to meet the demand and to bear the budgetary costs.

The relationship here between producer (DOE) and customer (DOD) is not the normal one encountered in commerce. In this case, the producer must seek to meet the customer's demand whether the funds available to do it are sufficient or not. We have not examined the basis for, or costs of, nuclear weapons—or the processes by which the demand is set—because the examination of such matters was beyond our purview. We perceive, however, that DOE is now obligated to produce nuclear materials and weapons through a decisionmaking process that may not reflect a full evaluation of the risks and costs of production—including health, safety, and environmental implications. Until now, when cuts in expenditures have been made to reflect available resources, health, safety, and environmental objectives have suffered rather than production. DOE, as indicated earlier, has become increasingly aware of the importance of accounting for risks and costs; it faces the challenge of assuring that the resulting calculus is included in the governmentwide decisionmaking process.

Setting Priorities

In order to meet its commitments, the Department must choose among a variety of projects that entail substantial budgetary costs: possible new facilities; expensive upgrades for health, safety, environmental, or production reasons; and remediation of the consequences of past practices. DOE has neither the technical capacity nor the budget to advance all the proposed activities at one time. Clearly, DOE must establish priorities in concert with DOD and the Congress.

Using Technical Strengths

Many activities in the complex are accomplished by using processes and practices that have not been substantially modified in over 40 years of operation. And much equipment is old. Although this state of affairs is not necessarily bad, the Department must be open to change where the incorporation of new technology is cost-effective and where it offers significant advantages in productivity and in the protection of health, safety, and the environment. Successful technical change requires the encouragement of the technical advancement of staff and a willingness to harness capacities both within and outside the complex to a greater degree than in the past.

Developing and Maintaining Competence

The operation of the complex presents serious technical challenges that demand the recruitment, training, and retention of qualified professional and technical staff. For some specialties, it is clear that the supply of appropriately qualified personnel is limited and insufficient to meet national needs; here, DOE may have to assume some responsibility for helping to improve the national capability. The complex also faces new challenges, such as remediation of contaminated sites, that require technical expertise not prevalent at the facilities in the past. Indeed, as in every human endeavor, the likelihood that DOE will adequately accomplish its mission ultimately depends on the technical quality of its staff and of contractor employees.

Changing the DOE Culture

The Secretary of Energy has observed that, in the past, the predominant focus of activities in the weapons complex has been on production. Now, however, consideration of health, safety, and environmental concerns must become an integral part of every facet of facility operations. Although management and other changes can assist, the achievement of this objective can occur only if a widespread and fundamental change takes place in the attitudes of federal and contractor employees and their respective institutions. An important component of the Department's effort to restore the public's confidence is the maintenance of competent staff that is motivated to perform its work safely in an environmentally sound manner, and that takes pride in doing so. All the Department's activities must be advanced with this end in mind.

2
Management

THE MANAGEMENT STRUCTURE

The facilities in the weapons complex are owned by the federal government and operated by private contractors selected and supervised by DOE. Management in the complex is the responsibility of the Assistant Secretary for Defense Programs (ASDP). Supervision of the contractors at each of the sites is provided by field offices, which are in the line management structure of the Office of Defense Programs (DP) (see Table 2.1). Field offices are of two kinds: operations offices and area offices. The principal official in an area office reports to an operations office, and the manager of the operations office in turn reports to officials at DOE headquarters. In the past, some managers of DOE field offices reported directly to the Undersecretary, whereas others reported to the ASDP.

The field offices receive direction from several entities at DOE headquarters, in addition to the Undersecretary and the ASDP. For example, direction is also given by the Assistant Secretary for Environment, Safety, and Health (ASEH) and the Assistant Secretary for Management and Administration, as well as by the General Counsel. At times the field offices have been challenged to respond to inconsistent directions from different offices at DOE headquarters. For example, a fundamental conflict exists between security and safety directives concerning the need to evacuate buildings rapidly in the event of an accident. Safety directives require emergency exits, yet some security directives require that personnel exit through manned security checkpoints.

The budgeting for and the coordination of the operations of the complex are, in the first instance, the responsibilities of the ASDP. However, because the costs of

TABLE 2.1 DOE Operations Offices and Facilities Subject to this Study in the Weapons Complex

Operations Office	Facility
Albuquerque Operations Office	
Management Support Division:	Sandia National Laboratory
Amarillo Area Office:	Pantex Plant
Dayton Area Office:	Mound Facility
Kansas City Area Office:	Kansas City Plant
Los Alamos Area Office:	Los Alamos National Laboratory
Pinellas Area Office:	Pinellas Plant
Idaho Operations Office	Idaho Chemical Processing Plant
Oak Ridge Operations Office	Ashtabula Plant
	Feed Materials Production Center
	Y-12 Plant
Richland Operations Office	Hanford Nuclear Reservation
Rocky Flats Operations Office	Rocky Flats Plant
San Francisco Operations Office	Lawrence Livermore National Laboratory
Savannah River Operations Office	Savannah River Site

operations are so substantial and because, in recent years, public and congressional scrutiny has become so intense, the Undersecretary and the Secretary have also become intimately involved in budget, priority setting, and planning activities.

Until recently, oversight to assure that operations comply with health, safety, and environmental and requirements and with other DOE orders (see the discussion of DOE orders below) has been the responsibility of the ASEH. Within the past several years, performance of this function was perceived to have become ineffective, and the Department made significant efforts to upgrade the scope, extent, and rigor of the oversight activities of the ASEH. Although the size of the Office of Environment, Safety, and Health (EH) staff increased as a result, neither this staff nor that of DP, which has line management authority, is large, especially when viewed in relation to the total number of people in the nuclear weapons complex. In FY 1989, DP and EH were authorized to have about 2,800 and 240 full-time equivalents, respectively, while the total number of employees among the operating contractors was about 64,000. Of the EH employees, about 85 percent are dedicated to DP. Because of personnel ceilings and difficulties in recruiting and retaining personnel with the necessary expertise, both the ASDP and ASEH have found it necessary to make extensive use of outside consultants to assist in the management function. The number of such contract employees helping DP and EH perform their functions in FY 1989 was about 220 and 300, respectively.

In the spring of 1989, the Secretary of Energy issued a notice (SEN-6-89, DOE 1989b) that the ASDP was to assume full responsibility for operational programs and activities related to health, safety, and environmental protection and that the

ASEH was to be relieved of responsibility for reactor and nonreactor nuclear facility safety. (See the discussion in the section "Changes" below.)

Contractors

Actual operation of the facilities in the weapons complex is performed by private contractors, and several contractors often perform different functions at the same site (see Table 2.2). Most of the technical expertise with regard to design, construction, and operations, as well as detailed knowledge of the facilities, resides with the contractors. If DOE seeks to address a health, safety, or environmental issue or to assess a problem that has arisen, it must place principal reliance on the on-site contractor. Maintaining the motivation and skills of the contractors is thus vital to the safe and proper operation of the complex.

In light of the special dependence of DOE on its contractors for technical and management skills, the relationship between contractors and the Department must be a partnership. The relationship is unusual in procurement situations: DOE cannot ordinarily provide detailed specifications that precisely define a contractor's obligations. Rather, DOE and the contractor must work together to confront challenges that, in part, will be defined during the contract term. Contracts typically provide the contractor with independent authority to manage its operations, including some leeway in spending available funds subject, of course, to audit. In matters of health, safety, and the environment, however, DOE retains full contractual authority to require the contractor to follow Department instructions.

The Department headquarters issues its formal operational and safety instructions to contractors principally in the form of DOE orders. Specifically, DOE has established or is seeking to establish a hierarchy of orders to describe safe practices in the design and operation of nuclear weapons facilities, as well as other facilities operated by or for the Department. Most government organizations require their contractors to meet agreed upon standards and specifications, but they do not issue specific directives after initiation of a contract. In contrast, DOE orders may become effective during the term of a contract and prescribe new requirements for conduct by DOE organizational units and contractors.

Award Fees

A Department contract typically provides that the contractor shall recover the cost of operations, plus a fee that is determined at least in part by the adequacy of the contractor's performance. To some contractors, the fee is not substantial, but it can still serve as an important incentive for plant managers. The operators of Los Alamos National Laboratory (LANL), Lawrence Livermore National Laboratory (LLNL), and Sandia National Laboratory (SNL) do not receive performance-based award fees. It appears that DOE has made little use of this mechanism for improving performance, since the fees awarded have not varied

TABLE 2.2 Management and Operations Contractors of Facilities Considered in this Study. Information provided by DOE.

Facility Management and Operations Contractor	Dates of Operation
Ashtabula Plant (1952)*	
Westinghouse Materials Co. of Ohio	1987–1992
Reactive Metals, Inc.	1963–1987
Bridgeport Brass Co.	1952–1963
Feed Materials Production Center (1951)	
Westinghouse Materials Co. of Ohio	1985–1992
National Lead of Ohio, Inc.	1951–1985
Hanford Chemical Separation Facilities (1945)	
Westinghouse Hanford Co.	1987–1992
Rockwell Hanford Corp.	1975–1987
Atlantic Richfield Hanford Co.	1967–1975
Isochem, Inc.	1964–1967
General Electric Corp.	1946–1964
E.I. du Pont de Nemours and Co.	1945–1946
Idaho Chemical Processing Plant (1951)	
Westinghouse Idaho Nuclear Co.	1984–1994
Exxon Nuclear Idaho Co., Inc.	1979–1984
Allied Chemical Corp.	1971–1978
Idaho Nuclear Co.	1966–1971
Phillips Petroleum	1953–1966
American Cyanamid	1951–1953
Kansas City Plant (1948)	
Allied-Signal Aerospace Co./KC Division (formerly Bendix Kansas City Division)	1948–1991
Lawrence Livermore National Laboratory (1952)	
University of California	1952–1992
Los Alamos National Laboratory (1943)	
University of California	1943–1992
Mound Plant (1947)	
EG&G Applied Technologies	1988–1993
Monsanto Chemical Co.	1947–1988
Pantex Plant (1951)	
Mason & Hanger-Silas Mason Co., Inc.	1956–1991
Proctor & Gamble	1951–1956
Pinellas Plant (1957)	
General Electric Corp.	1957–1993
Rocky Flats Plant (1951)	
EG&G Rocky Flats Corp.	1989–1993
Rockwell International	1975–1989
Dow Chemical	1951–1975

Table 2.2 continues

TABLE 2.2 Continued

Facility Management and Operations Contractor	Dates of Operation
Sandia National Laboratory (1948)	
AT&T (formerly Western Electric)	1949–1993
University of California	1948–1949
Savannah River Site (1953)	
Westinghouse Savannah River Co.	1989–1993
E. I. du Pont de Nemours and Co.	1953–1989
Y-12 Plant (1943)	
Martin Marietta Energy Systems, Inc.	1984–1994
Union Carbide Corp.	1947–1984
Tennessee Eastman Corp.	1943–1947

*Dates in parentheses indicate start of operation.

significantly over time. We recognize that some, perhaps most, contractors are motivated by more than the desire for profit—by a sense of public service, or the opportunity to develop or sharpen expertise that can be applied elsewhere, or the chance to maintain a presence in nuclear activities at a time when commercial nuclear opportunities are declining.

Contractor Turnover

In recent years the principal contractor at a number of sites has changed (see Table 2.2). When a new contractor assumes responsibility, the top management of a site may be replaced, but as a practical matter most of the existing staff are retained. The turnover among contractors thus does not reflect a drastic upheaval in contractor personnel, most of whom have been employed at their sites for many years. This pattern is a source of both benefits and costs. The facilities are manned by staff who are familiar with operations from long experience, but they are also accustomed to the old attitude that production automatically takes precedence over health, safety, and environmental goals.

The recent turnover in contractors is, evidently, in part a consequence of conclusions reached by corporate boards of directors that the management of DOE weapons facilities is not sufficiently rewarding. The public outcry over health, safety, and environmental problems has not made the contractor's function any more attractive. The problem is aggravated by considerations of liability. As DOE acknowledges the authority of state and local jurisdictions in matters of public health, safety, and environment and narrows the indemnification it offers to its contractors, it creates the prospect that a contractor might incur substantial fines for noncompliance. Thus, considerations of liability may serve to reduce the pool of potential contractors who are prepared to operate facilities in the complex.

In the past 2 years, two major contractors decided not to bid for continuation of their contracts. At the same time, a third firm has aggressively pursued contracts at several major sites. If some of the smaller facilities are closed (see Appendix B), this third firm may play an increasingly dominant role in managing the nuclear weapons complex. The trend is contrary to the generally understood governmental intent to diversify its contractor force to avoid excessive dependence on a single performer in areas of national security.

External Oversight

The Department of Energy has become aware, partly as a result of a report issued by the National Research Council (NRC 1987) concerning the defense production reactors, that the Department and its contractors have become too insular. In response the Secretary (DOE 1987) chartered an independent Advisory Committee on Nuclear Facility Safety (ACNFS), a committee of knowledgeable individuals largely from outside the complex. Participation on the ACNFS is a part-time activity for its members, but the ACNFS has already succeeded in reviewing activities at many of the nuclear weapons facilities and has submitted numerous findings and recommendations to the Secretary.

In 1988, Congress enacted legislation (Public Law 100-456, 102 Stat. 2076) to create a new federal agency to act as an external oversight body, the Defense Nuclear Facilities Safety Board (DNFSB). The DNFSB is to be a five-member board appointed by the President with the advice and consent of the Senate and supported by a staff of 100. The board is directed by statute to issue reports from time to time examining and assessing the safety of the weapons complex. The reports are to be public to the extent possible. The Secretary will be required to respond to the recommendations of the DNFSB. If the DNFSB determines that its recommendation relates to an imminent or severe threat to public health or safety, its recommendation and that of the Secretary will be submitted to the President for decision. At the time this report was written, the board had not yet become operational.

CHANGES

The new Secretary of Energy has stated his desire to revise the current management structure of the Department significantly. As a first step in the reorganization, the Secretary issued a notice (SEN-6-89, DOE 1989b) on May 19 that provides some indication of his intentions. The notice has several components:

• To establish unambiguous internal accountability for the compliance of operations in the nuclear weapons complex with health, safety, and environmental requirements, the Secretary has placed these line management responsibilities entirely on the ASDP. The ASEH was explicitly relieved of responsibility for developing and coordinating policy for nuclear reactors and nonreactor nuclear facility safety; this responsibility is being assigned to the Assistant Secretary for

Nuclear Energy (ASNE). The ASEH will continue to have responsibility for health, radiation safety, environmental protection, and worker safety. (Also see Tuck 1989.)

- The ACNFS will be directed to cease its examination of safety issues that are within the purview of the congressionally chartered DNFSB, once the Chairman of the DNFSB provides notice that that board is prepared to assume oversight responsibilities. The ACNFS will continue to exist, but it will focus on operations other than those being examined by the DNFSB.
- The manager of the Savannah River Operations Office was directed to report to the ASDP, rather than to the Undersecretary as in the past. This change was intended to clarify the responsibility of the ASDP for operations. At the same time, the Secretary created several new positions, relating to the restart of the Savannah River reactors, that also report to the ASDP.

The changes reflect a recognition by the Secretary that the management challenges facing DOE require careful reexamination of how the Department has operated in the past. Indeed, the order describing the changes stated that an "extensive review" of organizational structures and management practices was under way and that further announcements would be made.

In September, the Secretary described additional organizational changes affecting oversight of nuclear facility safety within the Department (Watkins, letter to J.F. Ahearne, ACNFS, September 1989; SEN-6A-89, DOE 1989c). He established offices within DP, the Office of Nuclear Energy (NE), and the Office of New Production Reactors that would provide independent checks on nuclear safety performance in the respective line organizations. The Nuclear Self-Assessment Offices will report directly to the senior program officials in the respective offices. In addition, he established a separate Office of Nuclear Safety with broad responsibilities to monitor and audit all aspects of nuclear safety in the Department, reporting directly to the Office of the Secretary.

We agree with the Secretary that line management under the ASDP should have undiluted responsibility for all aspects of the operations of the weapons complex, including safeguarding health, safety, and the environment. Line management now clearly has the obligation to satisfy multiple objectives, and production can no longer have priority over health, safety, and the environment. Yet, a process must still exist by which unavoidable conflicts can be confronted between production targets and health, safety, and environmental obligations in the face of limited resources of budgets, facilities, and personnel. As discussed later in this chapter, the Secretary's reorganization plan does not yet adequately address how these conflicting needs are to be reconciled.

AREAS FOR IMPROVEMENT

The many problems now confronting DOE in connection with the weapons complex are in large part the cumulative result of past management deficiencies.

We believe that restructuring should be guided by the following considerations:

1. Lines of authority and responsibility in the management and operation of the facilities should be clear, simple, and unambiguous.
2. Decisions about any issue should be made initially at the lowest level of management with the competence and authority to make them.
3. An internal safety oversight staff should be maintained with the power to raise issues to the next management level when unresolved conflicts arise.
4. Effective communications with contractors and among the different offices at each level in the management structure are essential.
5. Technical advice and assistance from outside the organization should be sought to gain fresh perspectives.
6. Constant attention should be paid to the maintenance and improvement of the technical capabilities and morale of personnel, whether federal or contractor employees, upon whom the effectiveness of the entire program depends.

These principles are rudimentary, but we perceive that the Department has not applied them consistently in the past.

Simple Lines of Authority

Conclusion *The weapons facilities are operated under a complex management structure with ambiguous lines of responsibility and authority.*

Although the ASDP has initial line responsibility for budget issues and for the overall management of the nuclear weapons complex, several managers of operations do not report directly to the ASDP, but instead report to the Undersecretary. In the case of the manager of the Savannah River Operations Office, this special line of reporting has been changed by SEN-6-89 (DOE 1989b), as noted above. The reality is that in the past the operations offices have not received much uniform central direction or control. This is obvious on even a casual visit to the sites.

The Secretary's notice is intended to clarify the responsibility of the ASDP as the responsible line manager for operation of the complex. Hereafter, the manager for the Savannah River Operations Office is to report to him, and presumably, this prefigures similar changes in the reporting responsibilities elsewhere in the complex.

The responsibilities of the various operations and area offices with respect to headquarters, each other, and the contractors are largely the product of history. The Albuquerque Operations Office, for example, has broad responsibilities. It supervises a full range of activities at a variety of weapons facilities across the country, so it cannot specialize its expertise or focus its attention on particular types of operations. The office also seems to take care of matters that might be considered DOE headquarters functions, such as evaluating and setting priorities for all requests for improvement of facilities across the entire nuclear weapons complex. Whether or not these appearances inhibit efficient management is unclear to us, but they deserve careful examination.

Recommendation *The Secretary should continue his efforts to simplify the line management structure for the complex, establishing clear and unambiguous lines of authority and responsibility.*

Decisionmaking Processes

Conclusion *Many decisions are now unnecessarily deferred by staff to higher management levels, sometimes creating delay and paralysis in decisionmaking.*

Although it is clear that upper management should be fully informed about controversial matters and obviously should have full discretion to revisit any decision made by subordinate decisionmakers, the system should encourage initial resolution at the lowest management level with the competence and authority to resolve a matter, subject to review if necessary. This is the only approach that can ensure that upper management levels are not swamped with unresolved matters, thereby increasing the likelihood of faulty or misinformed decisions. The system is too complex to be managed by just a few decisionmaking individuals.

The budget process provides a case in point. We were informed that all budget issues relating to environmental and safety issues are routinely referred to the ASDP, and often to the Undersecretary, for resolution. Although the sensitivity of such matters in the current climate might explain the reluctance of managers to make even tentative determinations of priorities, the fact remains that not all issues can or should be addressed at the highest levels in the Department. Indeed, the decisionmaking process should limit the issues that reach the upper levels of the Department to those that present only the most difficult judgmental or policy questions.

Of course, if staff at lower levels are to resolve issues in a fashion consistent with the objectives of upper management, management must provide clear guidelines for decisionmaking.

Recommendation *The Department should strengthen its management structure by delegating authority and responsibility for the initial resolution of issues to the lowest possible management levels, subject to clear guidance and support from upper management.*

Internal Oversight Structure

Conclusion *An oversight body internal to DOE but outside line management, such as the organization under the direction of the Assistant Secretary of*

Environment, Safety, and Health, is essential in ensuring the compliance of operations with health, safety, and environmental objectives.

As noted already, oversight of the line management of the complex has been provided in the past by the ASEH. Several studies, commencing with an internally chartered report (DOE 1981), have urged the Department to upgrade that office to assure that it can perform its function. Over the past several years, the Department has made continuing efforts to strengthen the EH organization through increased funding and substantially increased manpower. These efforts were noted with approval in previous reports by the National Research Council (1987, 1988b).

The Secretary stated in his May 1989 notice that the ASEH will no longer have responsibility for overseeing reactor and nonreactor nuclear safety within the weapons complex. The Secretary perceives the ASEH role to be diluting responsibility, and he therefore intends to give a single line manager, the ASDP, the responsibility for safety.

We believe that the Secretary has misperceived the function of the EH organization both to support DP with specialized expertise and to monitor the health, safety, and environmental performance of DP. We agree that the line management should have undiluted responsibility and authority for ensuring the safety of operations, as well as compliance with health and environmental requirements. The oversight function is not intended and should not be allowed to diminish that responsibility. Its purpose is to provide a second set of eyes to monitor activities and thereby to ensure that any deficiencies in decisions by the operational line management are reported and corrected before an accident or other adverse effect occurs. The currently popular maxim of "trust, but verify" applies.

The oversight function, if properly implemented, has an important role in the DOE management structure. As noted above, decisions should be made at the lowest line management level with competence if decisional gridlock is to be avoided. The oversight staff should monitor those decisions for their implications for health, safety, and environmental concerns. When it questions the operational decision, the oversight staff should be able to bring the matter to the next higher level of line management for resolution. If the system operates properly, oversight provides a mechanism for assuring that important issues—and a balanced and fair presentation of the facts—are brought up the management chain. In the absence of this tension between operations and oversight organizations, the system must rely solely on the strength of line management and an important element of redundancy in assuring safety is lost. Similar conclusions were reached in an evaluation of the management deficiencies in the National Aeronautics and Space Administration (NASA) that contributed to the space shuttle *Challenger* accident in January 1986 (Rogers 1986). In response, NASA found it appropriate to create an independent internal organization to oversee safety (National Academy of Public Administration 1986).

We recognize that the Secretary's actions in diminishing the role of EH may be explained by additional factors. First, the Department may have concluded that the current supply of technically capable personnel is insufficient to provide adequate staff for both line management and oversight. Indeed, the recent strengthening of the EH staff may in some instances have been at the expense of that of the DP. Moreover, the activities of EH may in some instances have been seen as intruding on the management prerogatives of the DP. There thus may well have been an impression within the Department that the EH had expanded beyond its appropriate bounds. Further, the ability to provide the operational line management with staff competent in health, safety, and environmental areas in competition with the DNFSB, the DOE contractors, and other private industrial organizations—as well as other agencies of government at all levels—may have led the Department to conclude that the ASEH could not retain or recruit the necessary staff to continue providing safety oversight.

Second, the Secretary's notice described the change in the ASEH oversight of safety in light of the pending establishment of the congressionally chartered DNFSB (see above). Because the DNFSB is to provide detailed scrutiny of line management to assure safe operations, the Department apparently has concluded that it is unnecessary to maintain an internal organization that would serve a duplicative purpose.

We agree that it is manifestly inefficient, perhaps even counterproductive, to establish multiple layers of redundant oversight, each with extensive staffs. Nonetheless, we believe that it is unwise to eliminate the responsibility for oversight within DOE that is separate from the DP line management. The DNFSB is not yet in place, and the new board should be given some opportunity to build its capabilities. More importantly, the DNFSB is designed to serve a function somewhat different from that of an internal safety oversight organization. Because it is a part of DOE's internal structure, the DOE safety organization can raise matters that affect safety internally for resolution within the line management organization in a way that DNFSB staff will not be able to do. Examples of such matters include the budget, allocation of resources, and maintenance—all of which entail decisions that affect safety but that would be beyond the reach of an outside agency. Indeed, many issues raised by independent internal safety personnel can and should be resolved at a local level without any involvement even of the DOE headquarters staff. The DNFSB, on the other hand, will not be part of the Department—it will speak formally to the Secretary through public announcements—and thus it will not be in a position to elevate issues within the Department's organization.

The difficulty experienced within the complex in providing for adequate maintenance illustrates the need for independent internal oversight. Managers have been and will continue to be under considerable pressure to meet production goals. In the past this circumstance has had a direct effect on the ability to conduct maintenance activities. For instance, the management of the Y-12 Plant told us that in the face of budgetary strictures, maintenance is the main victim. In 1984, about 25 percent of the plant was rated as being in poor physical condition or as having inadequate technology; the record has improved only slightly in the interim. We also received data at several sites about the growth of the backlog in responding to requests for maintenance. At the Idaho laboratory about 10 percent of backlog requests are more than 2 years old. At the time of our visit to Rocky Flats in March 1989, a considerable number of maintenance work orders designated as safety related were several months old.

The facilities at the complex are aging, and considerations of national security require that the physical plant remain in adequate condition. The specific funding level that should be allocated to maintenance or replacement of equipment is, in the aggregate, unclear. Rules of thumb applicable to industry may not be appropriate for the complex, and situations within the complex vary from one facility to another. It is clear, however, that maintenance has in general been shortchanged. In these circumstances of the long-standing and pervasive inability of line management to confront the problem, there is an essential obligation to maintain careful oversight. Officials independent of line management should identify the essential requirements for maintenance that affect health, safety, or the environment and should fight to assure appropriate resources for maintenance.

In sum, we conclude that an independent oversight body within the complex should be maintained to audit and monitor the DP for compliance at all levels and raise concerns with appropriate levels of decisionmakers. This oversight body need not be the existing EH organization, and oversight for safety, environment, and health issues need not necessarily be performed by a single entity. Although there is opportunity for organizational reform, the function of oversight should be maintained.

Whenever the cognizant oversight staff finds a lack of compliance with goals or applicable orders or regulations, or determines that designs, practices, or allocation of resources violate or threaten to violate reasonable standards at any level of management, they should attempt to ensure appropriate actions by DP line management. Failing agreement, the oversight staff should then have authority to

1. compel consideration of the matter to the next higher level of operational management in consultation with staff at that level; in cases of continued disagreement, it should be possible to continue this process up the management line eventually leading to the ASDP, the Undersecretary, or ultimately, to the Secretary; and
2. bring about a cessation of activity in cases of perceived imminent danger to the health and safety of workers at the facility or of individuals in neighboring communities.

As this report was being written in the early fall of 1989, it was unclear whether the recently announced Nuclear Self-Assessment Offices would eventually be able to perform the oversight functions we have outlined, because few details about their actual capabilities and operations were available. It appears to us, however, that the organizational solutions outlined thus far by DOE do not fully address the identified problems.

Because line management is to be given responsibility to oversee its own operations for safety, subject only to scrutiny by a newly created independent office (Office of Nuclear Safety) reporting directly to the Secretary, there exists no integrated system to elevate important safety issues beyond the ASDP.

In some cases, it may eventually happen that DOE's internal oversight program and DNFSB find themselves assuming unnecessarily duplicative roles. In these cases, the Department and the board should cooperate to determine just how each organization might adjust its functions to accommodate the other. It is premature now to determine the appropriate roles; the relationship must evolve with time.

Recommendation *The Department should maintain an internal oversight organization with the authority to seek resolution of issues within the line management structure.*

Communication

Directives from Headquarters

Conclusion *Problems exist in the development and content of communications from DOE headquarters to field offices and contractors.*

As discussed above, the vehicle by which DOE headquarters provides formally binding instructions on health, safety, and environmental performance to field offices and contractors is the series of DOE orders. We perceive problems in both the orders and the means by which they are developed.

DOE facilities, both within the weapons complex and elsewhere, differ significantly from each other. This diversity creates difficulties in the application of the orders. An order that provides concrete directions at one type of facility will not necessarily be appropriate at another. Ideally, orders should be so carefully drawn as to provide specific guidance and, at the same time, be flexible enough to address appropriately the wide range of different facilities to which they are applicable. It might be easier to tailor the orders to specific facilities or types of facilities, for example, research laboratories or production plants, instead of requiring all facilities to meet the same requirements.

The task of developing appropriate orders is difficult at best. The shortage of qualified talent at DOE headquarters aggravates the problem, and attempting to supplement this talent by acquiring the services of ad hoc contractors is often only counterproductive. Such contractors cannot bring the necessary breadth of experience to the task.

The process by which draft orders are reviewed is coordinated by a division of

DOE's Office of Management and Administration. This organization is responsible for refereeing the review and comment process. However, it is not staffed with personnel having technical expertise in health, safety, and environmental issues. Further, in a recent effort to respond to an earlier recommendation of the National Research Council (1987) regarding the need to strengthen the system by which DOE orders are promulgated, the Department established expedited schedules. The time available for external review of health, safety, and environmental orders was shortened from the customary period of 6 weeks to 72 hours. Moreover, DOE field offices and contractors have told us that their comments on draft orders appear to have little effect. Although the changes have allowed DOE headquarters to update and issue further orders, they have not had the benefit of a careful review process.

A case in point is a recently issued DOE order (5480.11) addressing radiation protection for occupational workers. The consensus of a contractor conference (Albuquerque, January 31-February 3, 1989) was that this order will not significantly reduce risk, that it will be expensive to implement, and that it is overly broad in its reach. Yet, as far as we have been able to determine, the results of the conference have had little if any effect on the order.

The process for maintaining and disseminating orders is itself antiquated. The entire series consists of 11 volumes that are maintained manually. An up-to-date index of the set is not available, and there is no cross-referencing system to identify orders pertaining to a particular subject or applying to a particular facility. DOE is currently investigating how to make the orders available on a computerized data base and should pursue this effort vigorously. Moreover, the process continues to lack a formal mechanism for reviewing orders on a periodic basis to determine whether there is need to prune unneeded or superseded orders or to revise current ones to bring them up to date with currently accepted standards and practices.

Operations offices are responsible for providing direction specific to the facilities within their purview based on orders issued by DOE headquarters. In general, however, the operations offices do not provide additional specialized direction, with the result that disparate facilities are governed by the same instructions. Further, when operations offices do provide tailored directions, no formal mechanism exists to ensure that the intention of the underlying order is met.

Recommendation *DOE should reform its system for preparing and promulgating its orders to address the deficiencies identified above.*

Exchange of Information within the Complex

Conclusion *Communications among organizations that confront common problems and efforts to focus the resources of the complex on finding solutions are inadequate.*

Many of the facilities in the weapons complex confront similar challenges in assuring safe and environmentally sound operations. Thus communication within the complex can help to ensure that each facility obtains the benefit of lessons learned at other plants and that resources are not needlessly spent with a variety of contractors independently seeking solutions to common problems. Moreover, communication can identify problems that, though only an irritant at any one facility, may be significant in aggregate across the complex. The full resources of the complex can then be focused on resolving such problems swiftly and efficiently.

We found that many of the contractors and the associated operations offices operated independently, with insufficient awareness of the existence of similar problems elsewhere in the complex. False alarms from the alpha continuous air monitors (alpha-CAMs) provide an illuminating example. These devices monitor for alpha particles, which indicate the presence of certain radioactive materials in the air, including plutonium. A number of facilities found that operations were frequently disrupted by alarms from the alpha-CAMs, most of them false alarms. It is customary when an alarm is triggered to clear the area immediately until technicians with protective apparatus can investigate. But the alarms are (or were) also sensitive to alpha emissions from radon, and normal background fluctuations of radon frequently triggered the alarms unnecessarily. Although the problem was occurring throughout the complex—interfering significantly with production at some locations—little or no effort was made to focus the full resources of the complex on solving it. Facilities worked on the problem independently, some without much progress. Personnel at the LANL, however, were particularly knowledgeable about the problem and conducted some research toward its resolution, but their insights were apparently not disseminated effectively to the other plants. Similarly, results of research on this problem conducted at other sites were not widely shared.

Although greater efforts at facilitating communication among the contractors and among DOE staff are needed, there are no simple prescriptions for accomplishing them. While decisions defining the responsibilities within the complex flow appropriately from the Secretary down, there must be encouragement of decentralized initiative by all levels of contractor and DOE staff. Such initiatives span the range from identification of health, safety, and environmental problems to the resolution of technical and management problems.

DOE is conducting topical conferences dealing with technical subjects of relevance throughout the complex. Examples include annual topical meetings on plutonium processing, modeling of environmental processes, operation of incinerators, and topics of interest to the medical directors of the respective facilities (see also Chapter 5). We received the strong impression that the opportunities for attending such conferences could be used more fully and that the attendees from each organization could share their information more effectively with their colleagues, so that they could learn as well as share. An electronic mail network is available to most, if not all, personnel working in the complex. These means of communications and others, both formal and informal, should be encouraged.

Recommendation *The Department should work harder to overcome the natural impediments to the flow of information among contractors and to facilitate communication among the contractors and among DOE staff.*

Independent Technical Advice

Conclusion *The Department is not aggressive enough in seeking the advice and counsel of experts from outside the weapons complex.*

For reasons that may stem in part from the legacy of secrecy that has surrounded the production of nuclear weapons, there has been a tendency by DOE and its contractors to look inward in confronting problems. The complex has been too insular and removed from the scrutiny of the public. Until recently, DOE has not sought external and independent review of operating or engineering practices in the complex. As a result, DOE has been urged to seek outside advice (e.g., NRC 1987, 1988b, and 1989a).

While not a substitute for having qualified people on the job, independent scrutiny can provide new insights and bring the benefit of outside knowledge to bear on the design and operation of plants and countless other matters. In addition, awareness of the involvement of respected outside authorities could help to restore public confidence in the work of the complex.

As noted earlier, the Department created the ACNFS to provide external peer review. The ACNFS has provided useful advice to the Department in connection with a wide range of issues over the short period of its existence. The Secretary has determined, however, that the role of the ACNFS should be supplanted by the congressionally chartered DNFSB with regard to those facilities that are to be within the jurisdiction of the new board. Although we concur that duplication of effort should be avoided, we believe that there may still be a continuing, albeit perhaps modified, role for the ACNFS in connection with the entirety of the weapons complex.

The ACNFS is composed of a range of individuals who have agreed to provide advice to the Department on a part-time basis. Its 15-person membership is otherwise engaged professionally with safety, environmental, and management matters outside the complex. The ACNFS thus brings a perspective to problems confronting the complex that derives from familiarity with other types of facilities, thus providing a cross-fertilization of views that the Department has lacked in the past. The DNFSB, which will have full-time commissioners and staff, cannot play entirely the same role: it will have fewer members; its function will be quasi regulatory rather than advisory; and over time, its focus and perspective will be defined by its involvement with the complex. Moreover, unlike the DNFSB, the ACNFS can be charged by the Secretary to examine particular issues of significance to the Department. The ACNFS and DNFSB are thus not necessarily equivalent, and there is benefit in retaining a role for the ACNFS in connection with an overview of the entirety of the complex.

Recommendation *The Department should aggressively seek outside advice, from the ACNFS and other sources, with regard to the many technical issues that it confronts in the operation of the weapons complex.*

Availability of Qualified Personnel

Conclusion *The effectiveness of the weapons complex in accomplishing its diverse, demanding tasks depends on the technical capabilities of DOE and contractor employees; qualified people trained in a number of relevant technical specialties are scarce.*

While it is widely recognized that upgrading of its facilities will be required if the weapons complex is to operate efficiently, safely, and without undue risks to human health or the environment, the performance of the complex ultimately depends on the technical quality and morale of the staff of DOE and its contractors. Expertise and skill can partially compensate for obsolescent facilities, but not even the newest and best facilities can be operated effectively without competent personnel.

The problem of attracting and retaining highly trained technical personnel is not unique to the weapons complex, but for understandable reasons the problem is aggravated here. Neither nuclear power nor nuclear weapons enjoy a favorable image in the eye of the public, and the numbers and qualifications of people completing studies in nuclear engineering or related fields or embarking on professional careers in the nuclear community have diminished markedly in recent years (e.g., NRC 1989b). Thus the pool of technically qualified personnel is small and shrinking. At the same time, the need for specialized expertise at the nuclear weapons complex is extensive and expanding. The fields in which the complex's needs are most acute include radiation effects, health physics, nuclear criticality, seismic analysis, environmental engineering, environmental toxicology, hydrogeology, and occupational medicine.

Recruitment

The overriding issue relating to the technical strength and vitality of the complex is, of course, the recruitment, training, and retention of the best available people. In this area, the recruiting prospects of the public sector are at a distinct disadvantage when compared with those of the private component of the system. The government's hiring process is cumbersome and time-consuming. The salary structure available to attract and retain qualified employees is severely limited and not competitive with private industry in either actual salary or fringe benefits.

Even if the special impediments could be ignored, the highly trained work force from which DOE might recruit essential employees is not large. The DOE contractors, other private sector firms, and universities, which have been dealing with these issues for a long time, are all in competition with the government in this arena. The net consequence has been a serious loss of expertise at DOE headquarters and in its field offices.

Moreover, often when their on-the-job training begins to make new DOE technical employees effective, they become prime recruiting targets for the contractors and other segments of private industry. The effectiveness of DOE headquarters has been further diluted by competition for personnel from other agencies, such as the U.S. Environmental Protection Agency (EPA) and the Nuclear Regulatory Commission, which also require personnel with similar expertise. All these circumstances have generated a virtual crisis in the shortage of people with technical expertise in fields related to the environment, health, and safety—a shortage that will be felt even more when the DNFSB is organized.

Within the contractor complex, the technical strengths and personnel compositions of the laboratory and production components differ strikingly. Many of the best scientific minds in the country are in the national laboratories system, working in facilities that are second to none. This technical community is highly educated: more than half the technical staff have university degrees, most of them advanced degrees. The laboratories conduct a broad spectrum of research and development, including areas related to safety and environmental issues, and they are attractive to people interested in scientific research at the vanguard. The challenge for the labs may be in the maintenance of skills essential to the production complex, such as criticality safety, in light of the small pool of available people in such fields.

The production facilities, on the other hand, have evolved within an entirely different culture, and they confront different employment challenges. The level of education of the operating work force is principally at the high school graduate level. Operators generally are given a modest amount of classroom training on procedures; the bulk of the training regimen consists of subsequent on-the-job training under the guidance of a "senior" operator. Technical oversight is performed by shift engineers generally with undergraduate-level training. However, most of these engineers have to learn the health-, safety-, and environment-related skills on the job. The challenge for the production facilities is to attract and maintain, in the face of intense competition, staff with the necessary skills in the health, safety, and environmental areas.

Finally, both private and public sector employers struggle with the painfully slow process of getting the required security clearances for new employees. This report is not the place to compare the benefits of security clearances and other measures to protect secrecy with the costs these measures impose on progress, and we do not have a ready solution to recommend. We are, however, convinced that additional resources should be expended to get proficient staff on the job faster. In particular, funds devoted to increasing the number of people performing investigations required for security clearances would be more than offset by the increased productivity of the individuals who are cleared more expeditiously as a result.

Recommendation *The Department should increase efforts to recruit highly competent technical personnel for all levels of its organization. Efforts should be made to speed the security clearance process. The Department should also establish the conditions necessary to retain personnel by providing them with opportunities for challenging assignments, participation in the decisionmaking process, and professional advancement.*

Training

DOE places little emphasis on training its employees working in the complex. The extent and quality of the contractors' training programs within the weapons complex vary greatly from exemplary to inadequate. One of the better contractor programs appears to be at the Savannah River Site (SRS). Courses are taught by plant personnel who are assigned to full-time training for a set period, after which they are transferred back into their previous positions. The rotation ensures that the course content and the instructors remain current. Similarly, trainees attend classes full time for the duration of their training assignment. The SRS program benefits greatly from an outstanding training facility (Building 705H), which contains, in addition to classrooms, laboratories with numerous mock-ups of facilities and computer simulations of both production and maintenance operations that allow realistic hands-on training without the interruptions and hazards of on-the-job training. The DOE training program at Savannah River has barely started.

In contrast, the program at the PUREX (plutonium-uranium extraction) Plant at the Hanford Nuclear Reservation appears to be poorly organized. It is oriented strongly toward on-the-job training with a minimum of classroom instruction, somewhat analogous to an apprenticeship program in a craft, although it is shorter and more job specific. The complexities and hazards of processing nuclear materials make it mandatory that production workers have some understanding of the theory of the processes, in addition to purely mechanical on-the-job aspects.

Programs at Rocky Flats and the Idaho Chemical Processing Plant (ICPP) suffer from a similar underemphasis on classroom instruction that needs to be rectified; fortunately, there are already encouraging signs at both plants. Rocky Flats has a new training facility and is apparently upgrading its training program.

And Westinghouse Idaho Nuclear Co., operator of ICPP, has adopted a fifth-shift policy to allow more time for training, a policy also being instituted at SRS.

While training can generate qualified personnel for the routine production and administrative tasks required for the safe operation of the complex, no training can compensate for the shortage of advanced scientific and technical talent available to the complex. There is no simple cure for the ills we have enumerated: they are part of a national problem.

Recommendation *DOE should require each major contractor within the complex to implement a strong training program with qualified instructors, adequate classroom sessions on theory, state-of-the-art mock-ups and computer simulations for hands-on experience, and where necessary, a fifth-shift schedule to allow adequate time for training. DOE should also place increased emphasis on training its own personnel.*

3
Environment

INTRODUCTION

In no other area has DOE been subject to a greater degree of scrutiny and criticism regarding its management of the weapons complex than in its activities related to the environment. Environmental issues in the complex have been the subject of numerous congressional inquiries, lawsuits, federal investigations, and a variety of activities by interest groups. The level of attention reflects intense concerns about environmental hazards associated with the facilities and a strong distrust of DOE in this regard based on past performance.

Background

From the time of the establishment of the Atomic Energy Commission (AEC) in 1947 until the legal ruling in *LEAF* v. *Hodel* in 1984 (586 F. Supp. 1163, E.D. Tenn. 1984), the operating agencies (AEC, then Energy Research and Development Administration, and now DOE) were perceived to have had sole responsibility to define programs to protect human health and the environment within the weapons complex. The waste management and environmental practices of the complex were born in the wartime atmosphere of urgency that understandably put a high priority on production but with little attention to the environment.

Initially practices relating to the disposal of chemical wastes were similar to those of other industries at the time. Wastes were placed in unlined and unprotected trenches; oils and organic solvents were poured into open standpipes in the soil; contaminated cooling water was deposited directly on the ground; unlined ponds that served as infiltration basins were used for the disposal of various liquid waste streams. Over the years, however, as other industries improved their waste management practices with the advent of environmental regulations, practices at weapons facilities changed comparatively slowly. Moreover, the environmental practices at the facilities were kept from public view under the shroud of secrecy that cloaked the complex from the early days of operations.

In 1984, DOE's self-regulated role was challenged by the lawsuit *LEAF* v. *Hodel*. The court found DOE to be in violation of the Clean Water Act (CWA) and the Resource Conservation and Recovery Act (RCRA) at the Y-12 Plant in Oak Ridge, Tennessee. With this legal ruling, the Department acknowledged the applicability of federal environmental laws, as well as certain state and local laws, to its weapons production activities. These laws generally provide for the regulation of air and water pollution and of the disposal of hazardous wastes, and they require the remediation of uncontained hazardous waste dumps. The radioactive component of wastes is not governed by RCRA, although the hazardous component of mixed wastes is. Management and disposal of the radioactive component of wastes is the responsibility of DOE under the Atomic Energy Act. (Final disposal of both commercial and high-level wastes is also regulated by the Nuclear Regulatory Commission.) Table 3.1 provides a list of major federal legislation affecting DOE's environmental programs.

The establishment of environmental standards for the weapons complex in principle requires the determination of the appropriate balance of risks and benefits arising from the operation of the facilities. It is by no means obvious that the balance for the weapons facilities is necessarily the same as that for industrial activities. Thus, at least in principle, a set of standards different from those applied to the commercial sector (perhaps more stringent, perhaps less) might be justified for application to the weapons complex. However, in the Five-Year Plan (see Recent Initiatives below), DOE, with the encouragement of Congress, has stated that the complex should operate in compliance with the environmental laws and standards that are generally applicable to industrial operations. This decision no doubt reflects a political judgment in which the Department, other federal agencies, Congress, the states, Indian tribes, and local communities have interests. The committee has not evaluated this judgment. We recognize, however, that the decision defines the agenda for DOE's activities related to the environment, and we have approached our task under the assumption that it will continue to do so.

With the Department's acknowledgment of the applicability of environmental laws, DOE set about performing the daunting task of bringing its facilities into compliance. Today, many former practices for disposing of wastes have been replaced with more modern and environmentally sound technologies. But the Department's initial efforts lacked an integrated plan and strong commitment from upper levels of management. More than five years after the *LEAF* v. *Hodel* decision and after many different planning exercises, the Department continues to be out of compliance in management of some of its current waste streams and is

TABLE 3.1 Major Federal Legislation Affecting Environmental Protection Programs

Atomic Energy Act
Federal Insecticide, Fungicide, and Rodenticide Act
Fish and Wildlife Coordination Act
National Enviromental Policy Act
Clean Air Act
Clean Water Act
Safe Drinking Water Act
Endangered Species Act
Resource Conservation and Recovery Act
Hazardous and Solid Waste Amendments
Toxic Substance Control Act
Comprehensive Environmental Response, Compensation, and Liability Act
Superfund Amendments and Reorganization Act
Emergency Planning and Community Right-To-Know Act

Source: SRS 1989.

lagging in its efforts to clean up previously contaminated sites. Dissatisfaction with its progress led several state environmental agencies, as well as EPA, to seek enforceable compliance agreements with DOE.

Recent Initiatives

The current Secretary of Energy has stated that the Department will reorder its priorities to put health, safety, and environmental issues on an even footing with production. With respect to environmental issues, this shift in emphasis has included a commitment to environmental restoration, compliance with applicable environmental regulations, and more cooperative relations with host states and regulatory agencies.

The cornerstone of DOE's recent initiative in this area is its *Environmental Restoration and Waste Management: Five-Year Plan*, issued in August (DOE 1989a), which is intended as a dynamic plan that will be updated annually. The document sets out the Department's plans for the next five years for coming into compliance and cleaning up contaminated sites. It describes how the Department intends to set priorities and establishes a goal to have completed the restoration of contaminated sites in 30 years. It anticipates spending $6.8 billion on environmental restoration activities over the next five years.

The Five-Year Plan makes an appropriate and useful distinction between environmental restoration required to remediate past practices and waste management operations stemming from current and future operations. These two aspects are reasonably distinct from one another and will require a balance in the efforts and resources of DOE. Additionally, the plan addresses the corrective actions that are required to bring facilities into compliance with existing environmental regulations and calls for applied research and development that is needed in support of restoration and waste management. This aggressive commitment to change, along with the comprehensive plan for action, is welcome and deserving of considerable praise.

ENVIRONMENTAL CONTAMINATION

Conclusion *Environmental contamination exists throughout the weapons complex. Addressing these contaminated sites is an immense task that will be costly and time-consuming.*

Virtually every facility in the weapons complex has some amount of environmental contamination within its boundaries, while many also have some contamination outside the boundaries. The severity of contamination is highly variable, with some sites containing very high concentrations of one or more contaminants and others containing rather small amounts. To date, more than 3,200 sites have been identified as having some soil contamination, or groundwater contamination, or both. Examples include the presence of cesium-137 in streams and wetlands at the Savannah River Site, plutonium-239 in soil at the Rocky Flats Plant, fission products in shrublands at the Hanford Reservation, and mercury in freshwater ecosystems at the Oak Ridge Y-12 Plant. Measurable groundwater contamination, usually involving organic solvents such as trichloroethylene, has been identified at several facilities including the Lawrence Livermore National Laboratory, the Idaho National Engineering Laboratory, the Rocky Flats Plant, the Hanford Reservation, and the Savannah River Site. Although this contamination is not likely to cause measurable disruption of ecological systems, there is concern that off-site groundwater movement could result in the ingestion of contaminants by people through drinking of well water or consumption of crops irrigated with well water. There is no doubt that the contamination is pervasive.

Our work did not include the review of risk assessments associated with the contamination. We therefore are unable to comment on the magnitude of the impact of contamination at the nuclear weapons facilities. It is the responsibility of DOE and the regulators to conduct such assessments. We note, however, that where contaminants are isolated from population centers by distance and natural barriers, there is reason to believe that the immediate threats to health are slight. Furthermore, we believe in general that ecological impacts of contamination—that is, the effects on systems of native plants and animals—have been less than the impacts resulting from construction of roads, buildings, and many other human activities at the sites. Nonetheless, the potential long-term risks to human health and the environment associated with environmental contamination at the sites pose difficult problems. The long-lived nature of some of the contaminants coupled with uncertainties surrounding the response of humans to low levels of chronic exposure to a variety of contaminants limit our ability to determine the effects. It is clear that the cleanup at the weapons facilities will be extensive.

Recommendation *In the Five-Year Plan released in August 1989, DOE outlines plans for an approach to environmental cleanup and, for the first time, commits to cleaning up sites in accordance with applicable laws and regulations. It is imperative that DOE proceed apace with remedial actions and that the Department, Congress, and the nation stand firmly by the commitment to clean up.*

SETTING STANDARDS AND PRIORITIES ACROSS THE COMPLEX

Conclusion *There is need to develop and apply a scientifically credible scheme to aid in making decisions about appropriate cleanup standards and priorities for performing remediation in the face of resource limitations.*

Implementing plans for environmental restoration will require DOE to resolve many complex questions with scientific as well as policy or social components. Among these, the most important at this time are those relating to acceptable levels of cleanup (or, conversely, what level of contamination may remain at a site after cleanup), how to clean up different sites, and which cleanup activities to undertake first. Not only are these difficult questions in themselves, but the answer to one will bear upon the answers to the others. Moreover, since cleanup activities are to be funded by public money, it rests with DOE to assure that its response and approach to these questions expend the public resources efficiently and cost effectively.

In the Five-Year Plan, DOE has identified two priority schemes it intends to use to guide decisionmaking regarding cleanup and allocation of resources across the complex. The first, the National Priority System (NPS) to be developed in consultation with EPA and other federal agencies, affected state and tribal governments, and other bodies, will be used to determine how resources should be allocated within and among corrective action, environmental restoration, and waste management operations projects. The second entails a plan to work with these same organizations to develop consensus-based, consistent regulatory standards for cleanup.

Risk-Based Cleanup Standards

In recent years, risk assessment has been used increasingly by regulators and other decisionmakers as a tool to bring scientific information into the decisionmaking process. One method for establishing standards is to determine the probability that an individual or population exposed to a given level of a contaminant will develop a fatal affliction because of the exposure over a lifetime and then set a standard based on what is considered an acceptable level of risk.

This probability, or risk, depends on knowledge about the dose-response relationship for contaminants and about the transport, dispersion, dilution, and decay of contaminants in the natural environment. It also depends on assumptions about pathways of exposure to human populations, for example, the amount of water consumed from a contaminated well or the ingestion of fish that have accumulated contaminants.

The dose to different individuals in a population will differ depending on where they live and on other kinds of lifetime experiences. One indicator of risk is the risk to a "maximally exposed individual," a hypothetical nearby resident who is assumed to be exposed to the largest possible amount of contamination through use of water from a contaminated aquifer or body of surface water, consumption of plants grown in contaminated soil, breathing air carrying contaminants, and so forth. Such an individual is assumed to dwell at the site for a lifetime. Another measure of risk is risk to a population. Population risk involves combining estimated risks to individuals with probabilities that a certain number of individuals will live within various distances from a site and be exposed to various amounts of contaminants.

Risk assessments can incorporate different measures of risk. In making a decision regarding how to manage risk (i.e., setting cleanup levels and priorities), however, many other important factors (e.g., economic costs, benefits, feasibility) will and should be involved. We use the term "risk-based analyses" below in a generic sense. We do not advocate any particular calculus; rather, we suggest that some consistent form or forms of risk analysis should play a role in determining standards and setting priorities. Our recommendations on the use of risk assessment presented below should not be read as a suggestion that risk assumes a role as a single determinant. We merely advocate the use of such analyses as a way to be consistent in using scientific knowledge to inform the decisionmaking process.

In a draft Five-Year Plan, DOE stated its commitment to "seek the establishment of technically sound, risk-based standards, which all can accept, to yield the greatest national progress toward DOE's compliance and cleanup commitments" (DOE 1989d). We encourage DOE in its effort to establish cleanup criteria on a risk basis.

There are, however, certain obligations that arise from a commitment to adopt risk-based standards. As the National Research Council's Board on Radioactive Waste Management (BRWM) commented in its review of the draft Five-Year Plan (NRC 1989a), as DOE moves through the transition from a self-regulated entity to an externally regulated one, it takes on the "obligation to participate, as an affected party, in rule-making and legislative initiatives that affect its mission." In this connection, there is a fundamental issue relating to the development of risk-based standards that requires DOE's immediate attention.

Any risk assessment involves (1) the evaluation of the toxic potential of the harmful agent(s) and (2) the estimation of the level of human exposure (or dose) to such agents. In the environmental setting, exposure to a hazardous contaminant will depend upon the ability of the agent to migrate in air, soil, or water, as well as on the proximity of an individual who may receive the dose. The exposure potential of an environmental contaminant will vary with site characteristics, including, but not limited to, soil type, precipitation, groundwater flow rates, and types of vegetative cover. A pathway analysis can be used to estimate how exposure will differ based on environmental setting and population parameters.

In general when risk-based standards are promulgated by EPA, a conservative or worst-case hypothesis exposure is assumed. Then a concentration of the contaminant in soil or water corresponding to what is considered to be an acceptable level of risk is determined for this case. Such determinations are considered risk management. This type of risk-based standard attains an acceptable risk level for the worst-case exposure, but it is overly protective in other circumstances. A standard that specifies a concentration in environmental media, for example, would actually result in disparate reductions in risk if applied at all sites without regard to issues such as the proximity of human populations to the source. In many situations, achievement of such a concentration standard is not necessary to assure the desired reduction in risk.

Another approach—the one favored by us—would be to establish an acceptable level of risk that contaminated sites should meet. The acceptability of risk is a value judgment, not a scientific one. DOE therefore must seek to establish an acceptable level of risk through a political process, although achieving consensus may be difficult. The specific cleanup level for a given situation would then be determined through the use of consistent risk assessment methodologies. We recognize that there are difficulties in this approach. Credible risk assessment methodologies can result in risk assessments that differ by orders of magnitude. Additionally, confidence in the assessment of the risks diminishes as projections are made further into the future. Nonetheless, tailoring cleanup requirements to specific sites seems to us to be the optimal approach, taking into account constraints in resources.

In the Five-Year Plan, DOE lays out the basic steps of the environmental restoration process required under RCRA and the Comprehensive Environmental Response, Compensation, and Liability Act (CERCLA or "Superfund"). These steps include preliminary assessment, inspection, characterization, evaluation of cleanup alternatives, cleanup action, and compliance. To the extent permissible by law, DOE should use risk-based methodologies to guide the Department through these phases and in setting priorities. In particular, it is imperative that risk assessment be used as a mechanism to bring scientific information into the decisionmaking, or risk management, process in the evaluation of cleanup alternatives.

The application of risk assessment methods in this context should take place on a site-by-site basis. In each case, characterization data should be used to estimate the risk to public health under several different courses of remediation ranging from no action to complete cleanup and employing different technologies, if alternative appropriate technologies are available. Risk assessments would be done using pathway analyses that model the fate and transport of contaminants in the environment and estimate the potential for exposure.

Although different sites will require different model parameters because of site-specific features, the methods used should be as uniform as possible. For example, the movement of chemicals through the environment should be analyzed in a manner consistent with the movement of radionuclides. The use of consistent risk assessment methods will not only ensure that the various alternatives for remediating a particular site are comparable, but they will also enable DOE to make intersite comparisons for priority setting with greater confidence.

In some cases, it may be found that the existing contamination does not warrant full-scale cleanup activities or that cleanup activities themselves may create greater risks. In such cases the prudent decision may be to institute only long-term monitoring to assure that the risk remains low or, if the risk is not acceptably low, to monitor until feasible remediation technology becomes available. In other cases, it may be found that the environmental impact of existing contamination is less than that normally associated with the construction of buildings and roads, and that to mount an engineered cleanup would cause greater environmental damage than leaving the waste in place.

Execution of the risk-based approach will be a demanding task. Ultimately, remediation decisions will have to be made in conjunction with the appropriate regulatory agencies. To come into compliance with legal requirements, DOE may be required to clean up to levels that are more than adequately protective of human health. It will, however, be to the Department's benefit to be aware of the risks and costs associated with its efforts so that it can defend a sensible, scientifically credible, and fiscally responsible approach.

Recommendation *The Department should seek to achieve site-specific cleanup standards. Consistent risk assessment methodologies should be used to bring scientific information into decisions regarding extent of cleanup, cleanup methodologies, and priorities for environmental restoration.*

National Priority System

As BRWM pointed out in its review of a preliminary draft of the Five-Year Plan (NRC 1989a), the development of DOE's NPS with the involvement of all affected parties is of crucial importance. The need for such a system is born out of the immensity of the cleanup task and recognition that it cannot be completed all at once. The NPS will guide DOE in approaching its task and in allocating resources. Although a final NPS has yet to be developed, DOE has proposed a four-tiered system to be used in the interim. The proposed system is as follows:

Priority 1: "Activities necessary to prevent near-term adverse impacts to workers, the public, or the environment," and "ongoing activities that, if terminated, could result in significant program and/or resource impacts."

Priority 2: "Activities required to meet the terms of agreements (in place or in negotiation) between DOE and local, State, and Federal agencies."

Priority 3: "Activities required for compliance with external environmental regulations that were not captured by Priority 1 or 2."

Priority 4: "Activities that are not required by regulation but would be desirable to do."

The distinction between priorities 2 and 3 appears to be based solely on the existence of compliance agreements and, as such, endows such agreements with an extraordinarily important role. DOE should view its responsibility to comply with all applicable environmental regulations as a single responsibility. DOE should acknowledge that it cannot come into compliance everywhere immediately and should place its initial efforts in areas where it can achieve the greatest reduction in risk with the available resources. The approach may require the renegotiation of some agreements already in place.

Recommendation *To the greatest extent practicable, DOE should incorporate risk assessment as a guiding principle in developing an NPS.*

Characterization of Contaminated Sites

Conclusion *Intensified sampling to describe the extent and nature of contamination, as well as hydrogeology and ecology, is necessary to guide cleanup, isolation, or restoration activities in a timely manner. Improved data management will assist in the retrieval and analysis of the massive amount of information collected.*

Characterization of the extent and nature of chemical and radioactive contamination at DOE sites is a prodigious task, but such characterization is essential before the nature and extent of any necessary cleanup activity can be determined. Identification and characterization of source terms, both past and present, and of contamination in soils, groundwater, surface water, and biota have already resulted in the collection of many thousands of samples. Sampling can be as simple as collecting vegetation or soil, as complex as setting up an air monitoring station, or as costly as drilling monitoring wells. Each sample may be analyzed for a host of constituents. For the results to be meaningful and useful, the samples must be collected and analyzed carefully, under strict quality assurance and quality control guidelines.

As with any undertaking, it is important to assess the costs of the effort relative to the anticipated benefits. Environmental sampling can be very costly, but the knowledge it supplies can lead to lower overall remediation costs. The extent of sampling should be tailored to the degree of risk presented by the site.

The problem of data management is obviously challenging, and it is made more complex because a large effort to collect data has been mounted over a relatively few years. As a result, the responsibility for collecting data has been fragmented at many, if not all, of the facilities. Maintaining consistent data bases in such circumstances can be difficult, but it must be done if the data base is to serve as a guide to decisionmaking regarding cleanup and further data collection.

The migration of contaminants in surface and subsurface media is complex. An understanding of the overall geology, hydrogeology, and land use at an installation, not merely the required measurements of concentrations of contaminants at each contaminated site, is required if informed risk assessments are to be performed. Indeed, analysis of the potential for migration of contaminants outside the boundaries of federal property demands such understanding.

Most work to date at the DOE installations has focused on thorough characterization of a relatively small number of contaminated sites. Nevertheless, an overall evaluation of environmental impact requires adoption of a broader, more consistent view applied across the complex.

Recommendation *Each installation should develop a comprehensive data base of environmental information, one that will allow the data to be accessed and used for a variety of purposes related to remediation of contaminated sites within the installation. The structure and content of the data base should be consistent across the complex. DOE should also insist that each installation develop a plan to acquire the data necessary to improve understanding of the installationwide geology, hydrogeology, and land use.*

WASTE MANAGEMENT

The Department's past production and waste management practices are easily identifiable as the source of the extraordinarily large cleanup and restoration projects facing the Department today. In looking to the future, it is important for the Department to learn from its past mistakes to prevent the occurrence of an analogous situation 40 years hence. A strong emphasis should be placed on improving current waste management practices and developing innovative waste management technologies for the future.

At each production site and research laboratory there are already in place waste management facilities providing some treatment and storage. These facilities provide for handling and movement of chemicals and radioactive materials followed by treatment to reduce toxicity, amount, or potential for exposure on dispersement. This general sequence of waste management occurs for radioactive wastes,

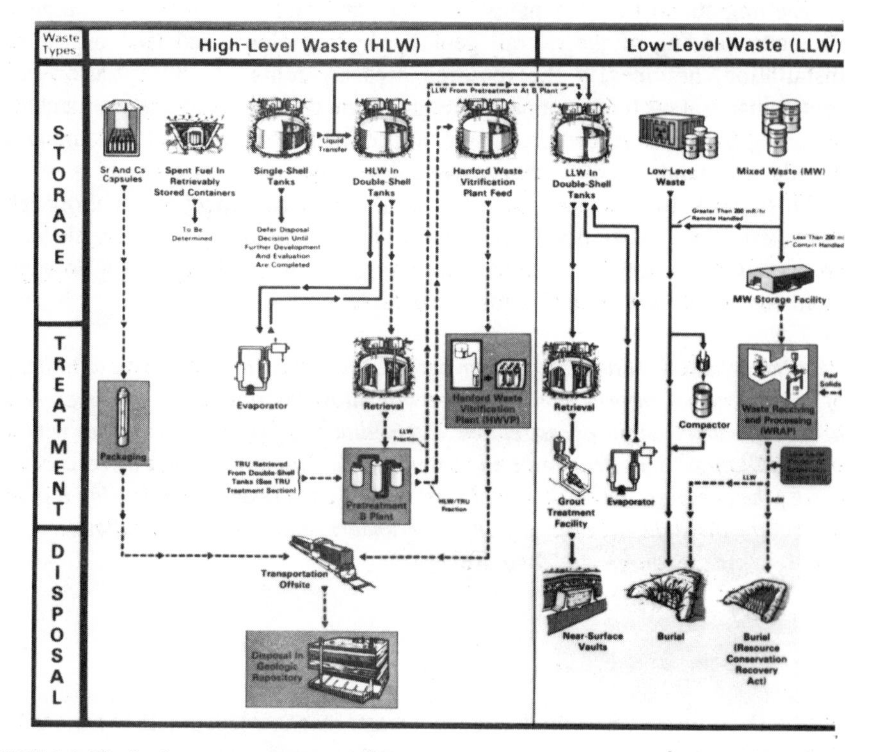

FIGURE 3.1 Hanford waste management plan.

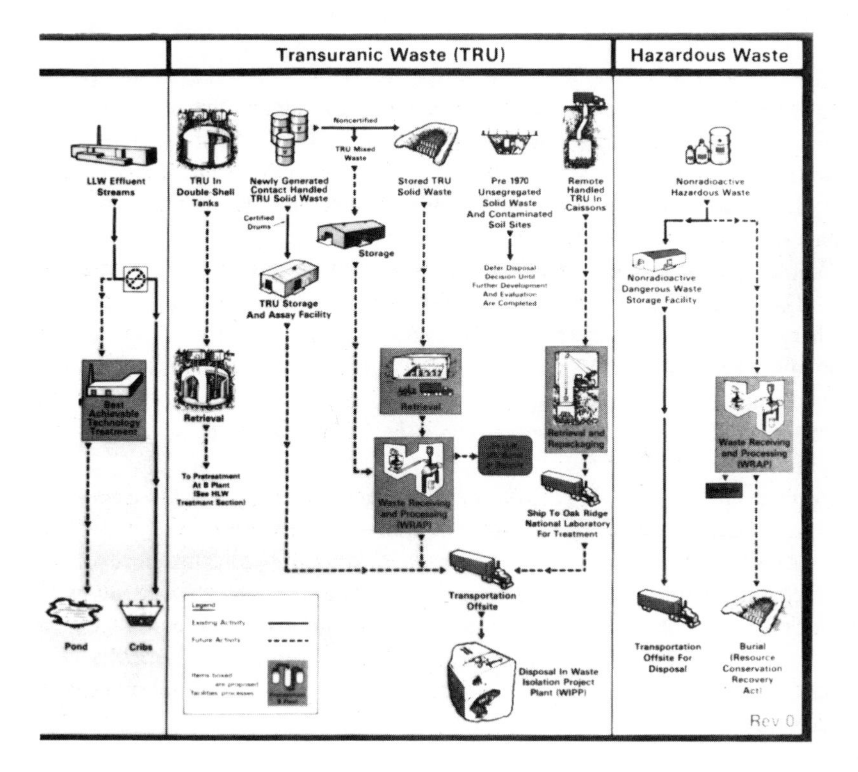
Transuranic Waste (TRU)
Hazardous Waste
LLW Effluent Streams
Best Achievable Technology Treatment
Pond
Cribs
TRU In Double-Shell Tanks
Retrieval
To Pretreatment At B Plant (See HLW Treatment Section)
Newly Generated Contact Handled TRU Solid Waste
Noncertified
TRU Mixed Waste
Certified Drums
TRU Storage And Assay Facility
Storage
Stored TRU Solid Waste
Retrieval
Waste Receiving and Processing (WRAP)
Pre 1970 Unsegregated Solid Waste And Contaminated Soil Sites
Defer Disposal Decision Until Further Development And Evaluation Are Completed
Remote Handled TRU In Caissons
Retrieval and Repackaging
Ship To Oak Ridge National Laboratory For Treatment
Transportation Offsite
Disposal In Waste Isolation Project Plant (WIPP)
Legend
Existing Activity
Future Activity
Nonradioactive Hazardous Waste
Nonradioactive Dangerous Waste Storage Facility
Waste Receiving and Processing (WRAP)
Transportation Offsite For Disposal
Burial (Resource Conservation Recovery Act)
Rev 0

hazardous and nonhazardous wastes, gaseous discharges, solid wastes, etc. These waste management sequences differ from facility to facility. An illustration of the current waste management strategy at Hanford, which addresses radioactive and other hazardous wastes, is shown in Figure 3.1. The ability of these waste management facilities to comply with evolving state and federal regulations for the various wastes that are produced is not uniform across the complex.

Conclusion *The Department needs an integrated long-range plan for waste management.*

The Department's Five-Year Plan is its first effort to develop an integrated plan that addresses waste management operations throughout the complex. The plan encompasses high-level, low-level, transuranic, and mixed radioactive wastes, as well as hazardous and sanitary wastes. The plan reflects the Department's priorities in this area, first, to comply with all applicable laws and, second, to reduce risks and minimize the generation of wastes.

We commend the Department for its intentions in this area. The strategy should be extended, however, to include nonhazardous liquid discharges, gaseous discharges, and thermal discharges, thus encompassing *all* process wastes and residuals. In addition, DOE should include wastes from decommissioning and decontamination activities in its planning. These wastes are expected to be substantial as the Department proceeds with the decontamination and dismantling of retired facilities.

Further, the Department needs to develop guidance as to how the Five-Year Plan will be implemented. In the course of this study, we observed that individual facilities often act independently of others that face similar problems. It behooves the Department to look at the entirety of its waste management operations to identify the more successful approaches. Although ultimately waste management options are likely to be assessed, adopted, and implemented locally, the establishment of an overall Department strategy would provide a useful framework for selection of alternatives.

Recommendation *The Department of Energy should build on its Five-Year Plan for waste management operations to include all production wastes and residuals, as well as wastes resulting from decommissioning and decontamination efforts. Additionally, the Five-Year Plan should be developed further to include plans for implementing the policies it set forth.*

Waste Minimization

Conclusion *The waste streams from different operations in the complex are diverse, can produce significant volumes, and may contain high concentrations of chemicals of concern. Reduction of wastes in the complex is essential.*

The management of industrial wastes presents an integrated or interconnected aggregation of choices, involving often conflicting technical and economic factors. Increasingly, industries in the United States are finding that principles of waste minimization provide a sound approach to these problems. The consideration of waste minimization varies now considerably from facility to facility within the complex. Most waste management programs place heavy emphasis on treatment and containment technologies, rather than on the often more cost-effective minimization at the source. Waste minimization principles could be applied profitably throughout the complex, not only in ongoing operations, but also to decommissioning and decontamination activities and to new initiatives as modernization proceeds. The application of an analytical framework based on waste minimization would give the Department an overall perspective on how the many residuals from its operations affect both the cost of operations and the environment.

Implementation of process modifications that result in less pollution per unit of production output often requires significant capital investment. Indeed, the restriction of capital outlays was frequently cited as a serious limitation for implementing change. Waste minimization programs at the facilities typically lack an economic framework or policy to provide a rationale for implementing cost-effective changes to reduce waste generation. Such a framework is necessary to determine whether, in fact, the heavy front-end costs offer significant long-term savings.

A potentially important, effective waste minimization effort is under way in the Office of Defense Waste Transportation and Management at DOE headquarters, and it involves representatives of most production facilities. The inclusion of source reduction goals in this program is relatively new to DOE. At this stage, we could not evaluate the effectiveness of this office. However, we view this effort as a hopeful sign that the Department is embarking on the right track.

As the Department begins to modernize the nuclear weapons complex, it should incorporate waste minimization concepts into the planning process at the outset. The economic, technical, environmental, safety, and health benefits of reduction in process losses and waste generation are far-reaching. The planning for the New Production Reactors (NPR), currently in progress, represents DOE's most significant effort in this area. The NPR program has already adopted a waste minimization approach with respect to radioactive wastes. Missing from these plans, however, are analyses related to minimizing hazardous and nonhazardous chemical wastes, solid wastes, low-level radioactive wastes, and other process residuals.

For a waste management policy that embraces waste minimization to be implemented successfully, a commitment to the effort must be made at all levels, from the operations personnel to the highest levels of management. Furthermore, because current environmental regulations focus on wastes that have already been generated, and not on source reduction, a commitment to waste minimization goes beyond a goal of compliance with state and federal regulations. Nonetheless, such a strategy offers significant long-term advantages.

Recommendation *The Department should develop and implement a framework for the effective and efficient application of waste minimization principles in all process and waste management contexts. DOE should make use of work done in large industries and academic institutions to develop an analytical framework for evaluating process and waste management alternatives and setting priorities.*

Process Development

Conclusion *Upgrading waste management operations to keep pace with future regulatory developments and to achieve waste minimization will require a sustained applied effort toward research and technology development and demonstration.*

We strongly support DOE's plans to improve available technologies and develop new ones with the aim of reducing costs and risk to workers and the public. DOE should also look outside the agency for nonradioactive waste treatment and disposal technologies. The commercial sector and other agencies of the federal government have supported research and development along these lines. For wastes that are unique to the DOE mission, however, DOE will have to support research and development activities.

Recommendation *The Department of Energy should sustain an applied research and technology development and demonstration effort in support of waste minimization and waste management activities. For wastes that are unique to the DOE mission, the Department must undertake wholly to support such efforts. In other areas, the Department should make use of technologies developed elsewhere and take advantage of research and development conducted under other auspices.*

ENVIRONMENTAL RESEARCH

An essential component of DOE's mission is the support of research on environmental and health effects of energy technologies. Over the years, DOE's commitment in these areas has waxed and waned. Today, when faced with the enormous task of addressing its extensive environmental contamination, upgrading waste management operations, and modernizing the complex, the need for more information on environmental effects and for the development of technologies to address these problems is great.

The Five-Year Plan recognizes the need for applied research and technology research and development to be conducted in support of environmental restoration goals. The BRWM has commented on how DOE could improve its proposed plan (NRC 1989a). In addition to applied research, however, it is imperative that DOE also support basic research directed toward the support of its cleanup goals and research mission.

Types of Research Needed

Conclusion *An increase in DOE-funded research will be required in order to meet needs in the areas of cleanup of contaminated soils and sediments and environmental restoration of damaged ecosystems.*

There is enormous diversity in the contaminated sites of the weapons complex in terms of area and depth of contamination, the type and chemical form of contamination, soil and biological characteristics, and other attributes. Current knowledge as to whether and how to remediate such contamination is growing, but limited. Because of these uncertainties, decisions concerning cleanup and restoration must be conservative and therefore may not be cost-effective.

If the risk-based approach we advocate is applied, the decision about whether to clean up a contaminated site must be guided by scientific knowledge concerning transport through environmental pathways and biological effects of the contaminants of concern. The mechanisms, pathways, and rates of contaminant movement from one medium to another must be understood. Moreover, it is necessary to determine how these processes and rates vary temporally, spatially, and with other factors such as soil characteristics. The more complete the information, the less the uncertainty with which risk estimates can be made. A good deal of information is currently available in this area, but site-specific, credible applications demand much additional information if large uncertainties are to be avoided.

Better understanding and quantification of the dose-response relationships are also needed to assess biological effects. For the weapons complex, determinations are complicated because, in many cases, two or more contaminants may be contributing simultaneously to biological stress. It thus would be helpful to understand the nature of the combined (possibly synergistic, possibly antagonistic) effects (NRC 1989c). Moreover, in addition to better knowledge of human toxicity, greater understanding of how chemicals and radionuclides affect critical species of plants and animals is required.

DOE's Office of Health and Environmental Research (OHER) now supports programs that deal with some of these needs. In our opinion, however, the magnitude and direction of current funding within OHER is neither sufficient nor properly focused. While environmental restoration technologies should be vigorously pursued, an equal effort should be placed on research to guide the need for and extent of cleanup. It does not make economic sense to spend massively to clean up contamination that currently is at levels that pose low levels of risk to humans and the environment.

In a similar vein, adequate funding now of waste minimization research, improved waste treatment and storage technologies, and environmental cleanup will also pay off in the long run by reducing the likelihood and magnitude of future contamination and necessary cleanup.

Recommendation *The Department of Energy should give high priority to research focused on the types of contaminated ecosystems in the DOE nuclear weapons complex and on the primary radioactive and chemical constituents that may require cleanup. The research should be clearly applicable to the development of models and risk-based standards for the cleanup and restoration of contaminated sites.*

The Role of DOE Measurement Laboratories

Conclusion *There is a serious erosion of technical expertise at the DOE measurement laboratories that could eventually have an adverse effect on the Department's interlaboratory comparison program and traceability of measurements to standards.*

Three DOE laboratories that specialize in the detection and measurement of radionuclides and internal and external exposures are a rich and unique source of technical expertise for the DOE weapons complex. They include the Radiological and Environmental Sciences Laboratory (RESL) at Idaho Falls, Idaho, the Environmental Measurements Laboratory (EML) in New York City, and the New Brunswick Laboratory at New Brunswick, Illinois. EML focuses on environmental measurements, the New Brunswick Laboratory is recognized as a standards facility, and RESL specializes in dosimetry. These facilities have contributed significantly to the DOE system in the research and development of new techniques for environmental measurement and dosimetry, in forming a central focus for interlaboratory comparison of measurement data, and in ensuring the traceability of standards for measurements made at DOE sites. The use of these facilities for the traceability of standards and interlaboratory comparisons of environmental and dosimetry measurements is crucial to the credibility of the Department's environmental and personal monitoring program. Each laboratory now employs experienced and highly qualified technical personnel, but difficulties in attracting and retaining young staff coupled with a lack of adequate support for the laboratories make the future of these facilities uncertain.

Recommendation *The role of the three DOE measurement laboratories should be carefully reviewed to ensure their optimal use and continuing quality.*

DOE'S ENVIRONMENTAL RESPONSIBILITY

In its relatively new role as an externally regulated entity, DOE must take on responsibilities in addition to legal compliance. Central to this role is the responsibility to communicate effectively with the regulators and the public about the risks associated with the Department's operations and how it plans to reduce them. The transition is extremely challenging in that it is taking place at a time when public confidence in the Department is low and concern about the environment is high. More than any other factors, public perception and satisfaction of the regulators will determine priorities for environmental issues for the future. It would strengthen the Department's credibility if it were to act as an active, open, and willing participant in this process.

Participation by the Public and by Local and State Officials in Environmental Programs

Conclusion *There is insufficient public and state and local government participation in the design and implementation of key environmental objectives, such as restoration of DOE sites and the assessment of risk to the public.*

The Five-Year Plan establishes DOE's commitment to working with regulators, the public, and others in establishing its priorities for environmental restoration and waste management. As pointed out by BRWM (NRC 1989a), however, DOE should be more explicit about its intentions. The National Research Council report, *Improving Risk Communication*, provides insight into how this can be done effectively (NRC 1989d).

Inadequate attempts to bring the public into the Department's decisionmaking process are often not the result of a lack of emphasis being given to environmental matters; rather, the cause is the extension of early philosophies of the weapons complex related to classification of operations and the serious lack of expertise and commitment to informing the public. At many sites, effective and comprehensive environmental monitoring and research programs are being carried out, but because of poor communication of results and the absence of public involvement, public understanding has not improved. At some of the sites, progress is being made through regular public meetings with technical personnel responsible for environmental programs, but little effort is made by key department and contractor management to participate in public meetings or to understand and address the concerns of the local public.

One example that may become a model for involving independent technicians and public officials is the Hanford Environmental Dose Reconstruction Project (Till, in press). This study is intended to reconstruct doses that individuals living in the vicinity of the reservation may have received from releases of radionuclides to the environment during the early years of operations. It is being directed by an independent steering panel, the Technical Steering Panel, while the study itself is being carried out by a major DOE contractor. DOE has clearly given the panel the authority to set objectives, establish priorities, and control the scientific quality of the research. The Technical Steering Panel sees its role as not only to approve and verify the data and models being used in the dose assessment but also to convey the results of the study to the public in a credible form once the work is completed. The second objective is likely to be the more difficult of the two tasks, and the panel is testing new ground in public involvement in a scientific study.

Although cooperation between host states and the Department seems to be improving at some sites, there continue to be serious differences regarding environmental quality, access to the sites, and what constitutes acceptable residual risk following cleanup. States must play a role in key environmental programs, including the design and implementation of monitoring programs for routine and accidental releases, and in establishing priorities for cleanup and restoration. Involvement of the public and state officials in environmental issues at the weapons complex facilities necessarily implies the responsibility of all parties to investigate and implement decisions from a sound base of scientific and technical information.

Recommendation *The Department and its contractors must significantly improve the involvement of the public and state officials in activities related to environmental issues at its sites. This participation must be actively and frequently sought in coming to terms with environmental issues.*

Ecological Value of DOE Lands

Conclusion *In general, DOE management does not appear to appreciate or understand the ecological value of the lands under its care.*

Several of the DOE installations, including the SRS, the Hanford Reservation, the INEL, the ORNL, and the Los Alamos National Laboratory (LANL), encompass large land holdings. All these installations have large and diverse natural areas that have been designated as National Environmental Research Parks (DOE 1985). In addition, smaller sites such as Rocky Flats and Site 300 at LLNL, although more heavily affected by operations, cover relatively pristine landscapes that are also significant in area.

The DOE sites were acquired 30 or more years ago when the lands were sparsely populated and were either unspoiled or in a condition to undergo natural ecological succession from old farmland. Protection from human disturbances and from intense development produced a diverse mosaic of natural ecosystems between the access roads, buildings, and waste disposal areas. In most cases today, the DOE land areas in pristine condition are far larger than the areas devoted to buildings, asphalt, concrete, gravel, or periodic disturbances. The natural areas include coniferous and hardwood forests, shrublands, grasslands, meadows, riparian thickets, cypress swamps, lakes, ponds, reservoirs, and streams. These ecosystems harbor a large diversity of native plants, fish, and wildlife, including several species classified as rare or endangered.

Much could be written about the ecological value of these sites, but in brief and among other things, they (1) serve as buffer zones for DOE activities within them; (2) naturally decontaminate air, water, and soil through geocycling processes over time and distance; (3) serve as breeding grounds and nurseries that help to replenish populations both on and off site; (4) act to prevent soil erosion and to recharge watersheds and aquifers; (5) provide harvestable natural renewable resources like timber; and (6) act as corridors for fish and wildlife migrations, as well as seed dispersal. By reason of their status under DOE's protection, these lands preclude urbanization, agricultural development, and destruction of historic archeological sites. Many DOE sites are excellent and unique places in which to conduct ecological research that could not be done on lands accessible to the public.

While these attributes are often not appreciated by DOE and contractor management, there are a few exceptions. The SRS is a model example of a DOE facility where an ecological culture has been adopted by many managers. Basic ecological research by the Savannah River Ecology Laboratory (SREL), as well as the Savannah River Laboratory, seems well appreciated by DOE and the site contractor management. Long-term, high-quality research along with popular writing and public lectures on the SRS wildlife and environment have obviously had a positive effect not only on plant management but also on public opinion. These activities thus have not only advanced science but they have also enhanced the independence and public credibility of DOE's environmental work.

Recommendation *Department of Energy and site contractor management should develop better awareness and appreciation for the ecological value of the lands under their stewardship. The on-site expertise in ecology and related environmental sciences should be maintained and strengthened. As a part of DOE's basic research program related to environmental remediation, the ecological resources of weapons complex installations should be made accessible to qualified researchers under the minimum constraints required to meet specific security interests.*

4
Safety

INTRODUCTION

From the perspective of conventional industrial safety, all the DOE nuclear facilities have excellent safety records. There are, however, other less conventional hazards at the weapons facilities, stemming from the handling of radioactive and fissionable materials, and these hazards are difficult to evaluate by the usual criteria of industrial safety. The most important of these is the exposure of personnel to radiation, both internally and externally. With few exceptions—one being nuclear medicine—the hazards of handling radioactive materials are unique to the nuclear industry.

Another unique hazard of fissile materials is the possibility of a criticality accident, i.e., the attainment of a self-sustaining nuclear reaction because of the inadvertent accumulation of too much plutonium or uranium-235 in an unfavorable configuration (see Appendix C). Criticality control has received strong emphasis at most sites, to good effect. Considering the large quantities of fissile material handled, the number of criticality incidents at processing facilities has been low (see Appendix C).

The Department has adopted and seeks to apply all the safety and health standards of the Occupational Safety and Health Administration (OSHA). In addressing radiation hazards, DOE has generally adopted the recommendations of the International Council on Radiation Protection (ICRP). However, DOE's enforcement of compliance with standards, from whatever source, is not consistent across the complex, and appears in some cases to be left largely up to the contractors.

This chapter includes a variety of observations and recommendations concerning the diversity of hazards arising from the operation of the complex. Clearly, some conditions are more serious than others. The committee believes that a particular sense of urgency is warranted in connection with fire safety, the handling of cyanide solutions, and the overreliance on respirators. We recommend prompt attention to these matters by DOE and its contractors.

INDUSTRIAL SAFETY

The weapons complex engages in many traditional industrial operations, such as metal fabrication, chemical processing, and electronic assembly. These operations can be evaluated by standards of conventional industrial safety. In 1986 the number of lost workday cases because of injury per 200,000 man-hours was 2.9 for all industry and 1.1 for the chemical industry, but only 1.0 for the DOE plants (National Safety Council 1987). This exemplary performance can be attributed to the strong emphasis placed on industrial safety by the DOE contractors.

The safety performance for radiation protection of personnel is well within the standards established by DOE Order 5480.11 and the ICRP guidelines. Radiation safety performance has improved considerably over the past 20 years, as indicated by the substantial reduction in the total dose received by employees with an exposure greater than 1 rem (see Figure 4.1). During this period, the number of employees in the complex has been relatively stable. As would be expected, the highest average exposures, within all DOE operations, are in the fields of fuel fabrication, reactor operations, fuel processing, nuclear components fabrication, and waste handling. For employees with a measurable exposure working in these areas, the average dose in 1987 ranged from 155 to 267 mrem, depending on the area (Pacific Northwest Laboratory 1989). Analogous exposure averages are slightly higher in the commercial nuclear electric power industry, but the comparison is complicated by the different operations and opportunities for exposure (e.g., steam generators) in the private sector.

In the following pages we note some examples of hazards that in our view deserve increased attention.

Inhalation of Radioactive Materials

Conclusion *Some facilities in the complex are contaminated. As a result, production workers need to wear respirators routinely as a means to prevent inhalation of radioactive contaminants.*

Plutonium, when inhaled, is an extremely toxic substance. Consequently, a central objective of industrial hygiene in the nuclear weapons complex is the prevention of exposure to respirable plutonium. Ideally, this objective is met by

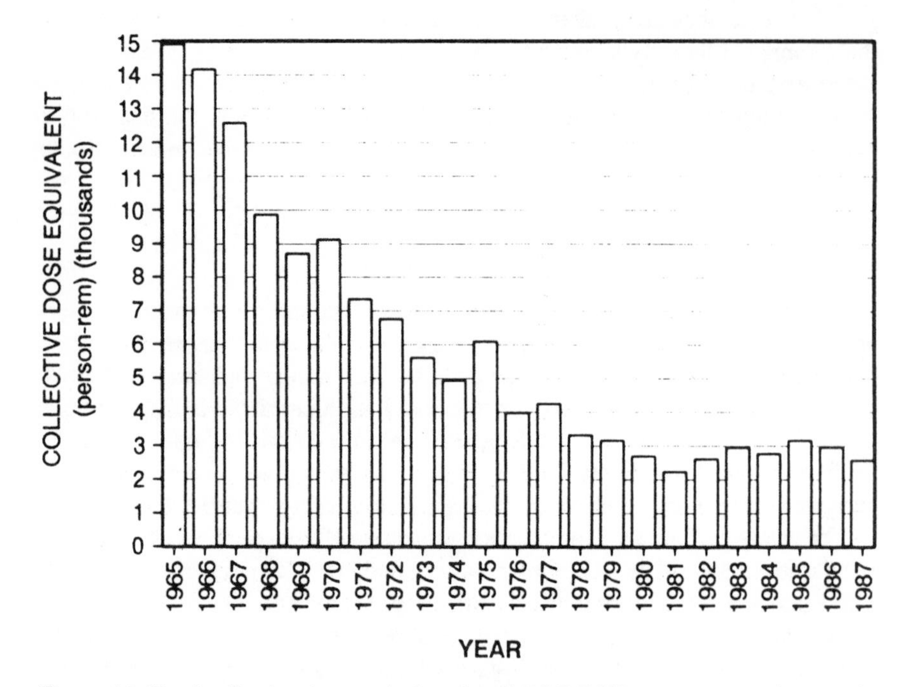

Figure 4.1 Total collective dose equivalent for all DOE-DOE contractor employees who received an exposure greater than 1 rem, 1965–1987.

adopting operating practices and contamination controls to avoid the need for breathing apparatus, such as respirators or supplied air, during routine production activities and most maintenance procedures.

Most sites within the complex have reasonably effective control programs, so that the use of respirators is reserved for emergency situations only, and supplied air is required only during certain maintenance procedures. In the event of an emergency, workers don respirators for protection while they leave an area. In some exceptionally well run facilities, such as Building TA-55 at LANL, it is deemed unnecessary for visitors to carry respirators.

At two sites that we visited, however, respirators were improperly used. In the plutonium production areas at the SRS, production workers are required to wear respirators whenever they are working in glove boxes as a precaution against pinhole leaks in gloves or other minor leaks. In our view, the mandatory use of respirators purely as a precautionary measure is unnecessary and counterproductive. Pinhole glove leaks can be greatly reduced, if not completely eliminated, by frequent inspection and radiation monitoring and by establishing a routine replacement schedule. Respirators are uncomfortable and impair employee alertness, efficiency, verbal communication, and morale. The risks from their routine use appear to us to outweigh the marginal protection they offer against potential minor radiation leaks.

The attitude toward radioactive contamination control at Rocky Flats is unique in the DOE complex. Some work areas are perpetually contaminated, and some production operations are conducted in a manner that makes contamination control virtually impossible. For many years there has existed at Rocky Flats a "respirator culture"—a feeling that as long as workers wear respirators, it is unnecessary to seek to maintain a contamination-free work area. The approach has led to sloppy operating practices.

The biggest problems at Rocky Flats are in Building 771. The high contamination levels there are not attributable solely to age, because the facility has been extensively refurbished over the past 6 years. One of the difficulties is the practice of conducting maintenance and production operations simultaneously; as a result, production workers frequently have to wear respirators for as much as 4 hours per shift. Even in the absence of maintenance activities, contamination is prevalent, and workers have to wear respirators for 2 or more hours per shift.

The overreliance on respirators has several negative consequences in addition to those listed above. Respirators place a strain on the lungs and increase fatigue. But perhaps their most serious disadvantage is that they engender a false sense of security—a feeling that, so long as a respirator is worn, there will be no radioactive inhalation problems. The fallacy of this conclusion is demonstrated by the experience at the ICPP at the INEL. In 1983 and 1984, the committed lung dose for workers at ICPP was more than 100 man-rem (see Figure 4.2), even though the wearing of full-face respirators in the contaminated work areas was required. In 1985 the lung dose dropped to 1.1 man-rem as a result of changes in work practices and by requiring the use of supplied air in place of respirators in certain operations. In our view, the pattern of use of respirators at Rocky Flats is an indication of the failure of production, maintenance, and housekeeping procedures. Maintaining an uncontaminated working environment is a more effective strategy than protecting workers in a contaminated environment.

Recommendation *The Department of Energy should discourage routine reliance on respirators in favor of engineered controls and operating practices that prevent contamination of the workplace. Respirators should be necessary only in emergency situations.*

Contamination in Ductwork

Conclusion *Sizable accumulations of plutonium exist in the exhaust ducts at some buildings that process the metal.*

An estimated 11 kg of plutonium has accumulated in the E-4 exhaust system, the 26-in. process vacuum piping system, and the stack manifold at Hanford's Plutonium Finishing Plant (Scientech Inc. 1989a). Some of the contamination is downstream of the high-efficiency particulate air (HEPA) filters, so that if the material were upset or dislodged, it could be released to the atmosphere. Kilogram

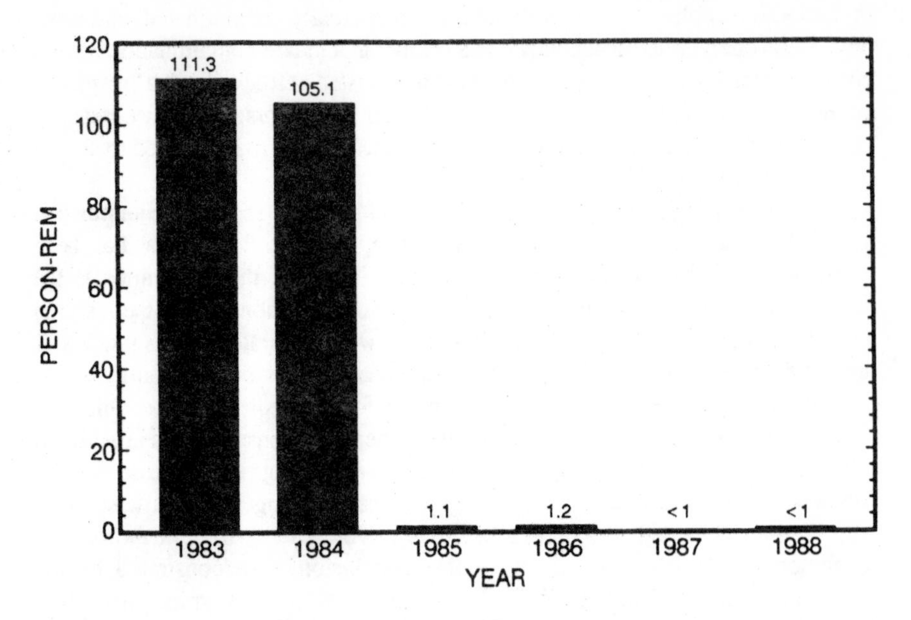

FIGURE 4.2 Committed lung dose equivalent; 50-year collective dose equivalent assigned to year of intake.

quantities of plutonium have also accumulated downstream of the HEPA prefilters in an exhaust duct of Building 771 at Rocky Flats (Scientech Inc. 1989b).

The hazards of accumulation are many, since any number of circumstances could cause a breach in integrity of the ducts. Earthquakes are an obvious dislodgment mechanism, as are releases resulting from corrosion, improperly performed maintenance operations, carelessness, or fire. Moreover, undesirable exposures of workers to neutrons may result even in the absence of a release, particularly if the ducts contain significant quantities of plutonium fluorides. Also, the threat of a criticality event makes it unwise to have accumulations of unknown quantities of plutonium in unknown configurations.

At Hanford an action plan has been developed, that calls for removal or cleanout of portions of the vacuum and exhaust systems in a pilot program commencing in FY 1989 and continuing through FY 1993. The program, which is expected to remove an estimated 6 kg of this plutonium, should be implemented soon and accelerated if possible. The cleanup should then be extended to the rest of the ventilation system.

The same problem may exist at other facilities as well. Even HEPA filters do not provide absolute barriers, and in any event, downstream contamination may occur when filters are changed. It thus seems prudent to assume that the ventilation systems in other plutonium processing buildings—the FB line at SRS, Building 332 at LLNL, and Building TA-55 at LANL, for example—may also contain plutonium in varying amounts. In addition, exhaust ducts in buildings processing uranium and beryllium could well contain unacceptable concentrations of these hazardous materials. A strategy for dealing with this potential contamination is needed.

Recommendation *The Department should develop and implement a plan to assess accumulations of plutonium, americium, uranium, and beryllium in the ventilation systems of relevant facilities and, in cases where significant quantities are found, institute cleanup or removal programs.*

Conventional Industrial Safety Practices

Conclusion *Some DOE contractors have indicated that criticality is their primary safety concern, and nuclear safety has been greatly emphasized. There are indications, however, of lack of adequate attention to conventional industrial safety practices.*

Some sites have a strong nuclear safety program, and the results are commendable. Most processing equipment is geometrically safe and physical constraints are provided to maintain safe spacing of fissionable material during transport and storage. Such stringent controls are lacking, however, in some areas involving conventional hazards. For instance, we observed the following conditions and practices at the Y-12 Plant. These observations, some of which are anecdotal, are based on circumstances—perhaps transient—that existed at the time of our visit. They are not intended as a condemnation of this site in particular, but rather as examples of the types of conditions that could exist—and should be eliminated—at all facilities.

- *Cyanide solutions are handled in a cavalier manner in the Plating Shop.* Gold plating operations with an acid cyanide bath are performed not in a full enclosure, but using only a fume hood or a horizontal duct just above the plating bath. This practice appears to be inadequate because cyanide salts in acid solutions are converted to hydrogen cyanide (HCN), a very toxic gas. In fact, because it is chemically such a weak acid, HCN is the primary cyanide species even in mildly alkaline solutions (up to pH 9). Its high solubility in water precludes a massive release of HCN gas into the atmosphere from the acid solutions commonly used in weapons production, but it is unwise to conclude that this reduces the need for adequate ventilation.
- *There are no high-efficiency particulate air filters on the exhaust system from the incinerator in the enriched uranium recovery facility (Building 9206).*

Instead, the exhaust system is fitted with bag filters. Regardless of whether the uranium release is within regulatory guidelines, this practice is contrary to the "as low as reasonably achievable" (ALARA) concept; emissions could be reduced by replacing the bag filters with HEPA filters.

Less serious were several observations indicative of poor housekeeping.

- *Storage practices were poor.* Cartons and bags of chemicals, some toxic and some leaking onto the floor, were stored on pallets in work areas and near high-traffic routes. In the loading area, large arrays of gas cylinders were stored without adequate anchoring, and some 55-gallon drums were stacked precariously.
- *Some floors were oily.* In the pressing area of the lithium facility, the footing was excessively slippery, particularly when shoe covers are worn. In the beryllium and depleted uranium machining areas, lathe coolant was spilling onto the floor. Rigid plastic housings similar to those on the enriched uranium lathes are needed.

Recommendation *While maintaining its commendable emphasis on nuclear safety, DOE and its contractors should reassess conventional safety programs and institute an upgrade to bring them on a par with nuclear safety.*

Sitewide Emergency Control Centers and Local Monitoring of Safety Systems

Conclusion *Sitewide emergency response plans do not effectively make use of knowledgeable personnel working within the various buildings. Monitoring of safety systems in buildings where a serious emergency might occur is inadequate.*

A number of sites have sitewide emergency control centers designed to respond to plant emergencies: the Rocky Flats Plant and ICPP at INEL are two examples. Such centers are necessary, but in some cases they are inadequate. For example, the ICPP center has the disadvantage of being near and downwind of an HF storage tank, so that it would be uninhabitable in the event of a major rupture or spill at the tank.

In general it is not possible for the staff of a sitewide emergency control center to have specialized knowledge of the operations and hazards in all the buildings at the site. Only persons permanently assigned to a building are likely to possess the necessary detailed information, such as current configurations and inventories. Therefore, in any building where an emergency might have serious immediate or long-term consequences, the emergency response team should be made up of people who work in that building. The teams should be linked to the site emergency control centers through procedures clearly understood by all concerned as laid down in the emergency response plan.

A related issue is the close monitoring of safety systems within each building to assure prompt response to abnormal conditions. In addition, operations should be contingent on the operational status of all essential safety systems. These systems might be measuring parameters such as ventilation flows and vacuum levels, the integrity of HEPA filters, air contamination, steam pressure, or temperature stability. Centralized monitoring is warranted to assure that all safety systems are operational. Within each building the personnel responsible for the localized monitoring of safety systems would be valuable additions to the in-building emergency response teams described above.

Recommendation *Any building where an emergency might have serious consequences should have an emergency response team that includes employees who are knowledgeable about that building. In addition, all essential safety systems within each building should be continually monitored to ensure that they are operating correctly.*

FIRE SAFETY

The fire protection program within the complex is multifaceted. It encompasses the following elements: safe operating procedures and administrative controls to minimize fire hazards; the design of structures and production systems to mitigate the effects of fire; the testing and maintenance of fire protection systems to assure their performance; and the organizing, equipping, and training of site fire departments to assure a prompt and effective response to any fires. Written guidance covering many aspects of this program is contained in DOE orders and other criteria supplemented by industry standards and the practices of contractors. The individuals responsible for implementing the program are a diverse group of knowledgeable and experienced fire protection specialists.

Conclusion *Fire protection within the complex is, to a significant degree, addressed on a site-specific basis, and decisions concerning individual issues are made by the local representatives of DOE or its contractors. Little coordination among sites was apparent, and an insignificant level of headquarters oversight to ensure consistency was evident. The inconsistency has resulted in a number of instances of fire safety issues being unevenly addressed across the complex or not addressed at all. This tendency has been aggravated in some cases by a lack of clear, explicit criteria from DOE concerning the design of fire protection features or the implementation of procedures to deal with fire protection issues unique to the weapons complex and not adequately encompassed by industry standards, such as the National Fire Protection Association (NFPA) Fire Codes. Despite these limitations, DOE and its contractors have achieved a number of noteworthy accomplishments. Among them are well-equipped site fire departments with a fleet of modern mobile apparatus and highly trained fire fighters. In addition DOE property losses due to fires are low.*

The Department's fire protection program criteria, as delineated in the various orders and other internal documents, provide acceptable statements of overall fire safety philosophy within the weapons complex. This general guidance is supplemented by reference to industry standards such as the NFPA Fire Codes. Unfortunately, industry codes do not adequately address several special requirements of the weapons complex that are not found in private industry. It is especially important that hazardous materials not be transported by fire-generated flow fields; carefully designed ventilation systems can help minimize this threat. Other special requirements include glove box fire protection, fire-safe ventilation in a radiation environment, emergency egress from secure areas, and the need for mobile fire apparatus at individual sites. The lack of special criteria has resulted in ad hoc approaches to fire protection across the complex.

The Department's fire protection program criteria require that fire suppression systems be installed in locations where a fire could cause damage to equipment that would interrupt process operations for longer than 6 months. At the Rocky Flats and Y-12 plants such "single-failure" areas were routinely protected by automatic or (on a limited basis) manual sprinkler systems. At the remaining sites we visited, there were locations of this type that were vulnerable to fire damage and not adequately protected.

We were also concerned about locations where a single fire could damage systems necessary for the safe operation of the production facility. There was no evidence that DOE or contractor fire protection criteria comprehensively address the provision of adequate fire protection for these locations. Moreover, there was no evidence that a systematic effort was being undertaken to identify such locations for future safety enhancements.

Fire protection design for ventilation systems within materials processing facilities varied widely among the sites we visited. The specific focus of our efforts was on filter plenum design. At Rocky Flats, the contractor has applied internally developed fire protection design criteria that are both explicit and conservative, featuring multiple stages of fire safety features. At other sites, more limited protection was observed. In some instances, only fire detectors were installed in return air plenums, in accordance with NFPA Standard No. 90 A. At other sites, fixed manual or automatic fire suppression systems were provided within filter plenums, depending on the size of the plenum.

With the exception of several recently constructed buildings, most of the structures and mechanical systems observed within the complex were erected and installed many years ago, and they were not designed to withstand the effects of the more severe earthquakes that might occur in their regions. Consequently, passive and active fire protection features may not be operable following a seismic event. Manual fire-fighting efforts would be hampered by the unavailability of water for hose streams and the distinct possibility of simultaneous multiple alarms from malfunctioning automatic systems. However, no contingency plans had been formulated by DOE or its contractors at any site we visited to respond to postseismic conditions.

At three sites—the Y-12 Plant, INEL, and SRS—we investigated the adequacy of fire department radio communications and found that, at all three, structural interference to communications was acknowledged as a problem. Specifically, within certain areas of some of the larger facilities the fire-fighting attack teams would not be able to communicate with each other or with supporting personnel because of the steel structural elements. At SRS, telephones were offered as a compensatory feature, but their viability in a smoke-filled environment could not be confirmed.

A related issue is the availability of a dedicated radio frequency for fire department use, which offers the advantage of no nonessential conversational "clutter" during a fire or medical response. The fire department at INEL has such a radio frequency, but at the Y-12 Plant, the fire department has to share its radio communications capacity with other site organizations.

Variations in Operational Approach

Fire protection systems designed to mitigate the consequences of a fire are not comprehensively or uniformly covered by operational safety requirements (OSRs) throughout the weapons complex. OSRs are facility-specific procedural requirements covering many different systems. For some critical mechanical and electrical fire safety systems, they mandate that alternative compensatory actions be available if those systems become inoperable. Based on interviews conducted with the fire protection staffs, it appeared that the Hanford Site has the most fire protection systems covered by OSRs. Most active fire protection features, such as fire detection and suppression systems, are covered at Hanford by OSRs. However, fire barriers, including fire doors and dampers necessary to restrict the spread of fire within a facility, are not covered by these requirements. The applicability of OSRs to fire protection features at other sites within the complex varies considerably; indeed, at the Y-12 Plant we were informed that no fire protection systems are covered by OSRs.

Based on interviews with DOE and contractor staff, we concluded that the fire protection organization's involvement (including that of the fire department) with emergency planning and preparedness at both Rocky Flats and SRS was well handled. The involvement included off-site organizations, DOE and contractor personnel, and the frequency of drills and simulation of accident conditions. At the Y-12 Plant, although drills were conducted, realistic conditions were not simulated and the drill frequency was lower than those at other sites.

Fire Protection Audits

The Department's contractors are responsible for performing periodic fire protection audits. Local DOE fire safety professionals also audit the performance of contractors. DOE headquarters appears to have a minimal role in this process. We investigated the adequacy of contractor fire protection audits, looking at frequency, comprehensiveness, report format, and the use of DOE criteria. We concluded that at most sites the audits were adequate, except for the coverage of critical single-failure points discussed above. At the Y-12 Plant, however, there was evidence that existing DOE fire protection criteria were not being used in the evaluation of site facilities. At Rocky Flats, in part because of the personnel shortage, audits were significantly less frequent and less detailed than those at other sites.

Personnel and Equipment

The organization, staffing, training, and equipment of the site fire departments were, with few exceptions, superior. Each site benefited from a fleet of modern mobile apparatus, including support vehicles equipped to deal with most contingencies. The fire fighters appeared to be motivated, and they had undergone extensive training. Our only criticisms concern the absence of criteria governing the selection of vehicle types, the siting of fire stations, and the determination of minimum personnel levels.

Personnel levels, for both fire protection engineers and fire fighters, are adequate for current needs at Hanford and the Y-12 Plant. At INEL and SRS, some vacancies in the contractor fire protection engineering staff have had an adverse impact on the fire protection program, reducing the frequency of periodic audits. Personnel shortfalls are most severe at Rocky Flats. No DOE fire protection engineer is available, and only one contractor fire protection engineering position is currently filled. Several additional positions are required to fully staff the site fire department.

Funding for modifications related to fire safety was adequate at most sites. Fire protection line items are in the budgets at the Y-12 Plant, INEL, and SRS, but not at Hanford or Rocky Flats, where some needed safety improvements were delayed because of insufficient funding.

Recommendation *DOE should develop specific engineering design criteria and administrative guidelines for fire safety for application to the special problems of the complex. These criteria and guidelines should benefit from input from the individual site fire protection staffs and allow for diversity of application depending on local conditions. DOE headquarters should more actively audit the sites to assure that criteria are being implemented in an effective manner to achieve a consistent level of fire safety throughout the complex.*

CRITICALITY SAFETY

Conclusion *Department of Energy contractors are generally providing effective criticality controls for operations with fissile materials. A shortage of criticality safety personnel exists, and the future of the one remaining facility available for training in criticality safety is uncertain.*

Current criticality safety practices in the various DOE contractor organizations are generally outgrowths of the control systems established in the 1960s, following the series of criticality accidents between 1958 and 1964. Details have evolved somewhat differently for each contractor, as is probably appropriate; but in these organizations, the basic criticality safety standards are being met. These standards were developed by the American Nuclear Society Standards Subcommittee 8 and through Consensus Committee N-16 (American Nuclear Society, 1975-85). (These documents also provide the bases for the Regulatory Guides of the Nuclear Regulatory Commission that address criticality safety concerns. The Nuclear Regulatory Commission relies more heavily on these standards than does DOE.) Morover, under the sponsorship of the Nuclear Criticality Technology and Safety Project funded by DOE, annual conferences provide opportunities for discussions of problems that have been encountered, current practices, and changes that have been proposed in standards and DOE orders related to criticality safety.

A concern recognized at most facilities was the difficulty of finding and training people for criticality safety assignments. Even to sustain the current efforts in ensuring criticality safety, DOE and its contractors will have to recruit and train personnel to produce experts in this highly specialized field. Training programs and facilities are obviously key to success in this area.

Many of those who have served in this activity over the past 25 years gained their experience through work in the several facilities that were conducting measurements on critical assemblies. Today only two facilities of this sort exist, and the one at Rocky Flats is dedicated to the solution of problems related to production at Rocky Flats. The Los Alamos Critical Assembly Facility (LACAF) is the only remaining general purpose facility, and its assembly machines provide measurements in support of other contractors, in addition to serving Los Alamos' needs. The facility is also used for giving hands-on experience to students attending the 2- and 5-day classes in criticality safety conducted by the Los Alamos Criticality Safety Group. One other contractor, Martin Marietta, is sending new and less experienced criticality safety personnel to LACAF to gain the perspective provided by work in such a facility.

Criticality safety practices today make great use of the large computer capabilities available in the complex. But such modeling efforts have their limits. Computer models must be carefully verified by experimental data. For example, it is important that statistical information arising from the experiments be correctly treated in the computer model. Moreover, wherever possible, computer-developed designs should be evaluated with experimental data. A revised broadly applicable set of such data has recently been published (Paxton and Pruvost, 1986) through the support of the DOE Office of Nuclear Safety, now defunct. These data are predominantly related to aqueous systems containing enriched uranium or plutonium, although such other data as exist are included.

For innovative designs and procedures, criticality safety evaluations must rely either on newly acquired data or on large and uneconomic safety margins. As an example, pyrochemical operations, such as the direct reduction of plutonium oxide to metal, are currently undertaken using small batches of fissile material. If the operation were to be scaled up, further experimental measurements on systems containing the fissile material and salts would facilitate process designs that are both safe and efficient. A capability to make the necessary measurements must be maintained, or the ability to achieve process efficiencies with new technology will be reduced.

In the absence of an organization like the former Office of Nuclear Safety, DOE has no focus for conducting criticality safety measurements important to all nuclear facilities. DOE policies state that criticality accidents must be prevented, but the concomitant support is not always provided. In the distant past, contractors had the flexibility to assign resources to needed activities, and it was during these times that most of the data regarding criticality that we rely on today were generated. The number of critical experiments performed today is only a small fraction of the number carried out 25 to 30 years ago. This is partially due to the wealth of accumulated data, but it is also attributable to the increased complexity of regulatory requirements, limited funding, lack of a clear assignment of responsibility within DOE, and the fact that most of the "easy" experiments have been done. One of the working groups of the Nuclear Criticality Technology and Safety Project has developed a prioritized list of criticality measurements to be performed (Brown 1987).

The Department of Energy has recently chartered the Nuclear Criticality Safety Program Committee, composed of program officers involved with criticality. DOE has done well to form such a committee to study the questions of where responsibility for criticality safety should be assigned and what criticality experiments are needed. This is a good start toward rationalizing the organization and program for criticality safety in the weapons complex.

Recommendation *The Department should continue its effort to develop and implement a coherent criticality safety program. DOE must alleviate the serious shortage of technical personnel in criticality safety through an enhanced training program.*

SEISMIC SAFETY

Conclusion *Over the past decade, seismic design criteria for new DOE facilities have been consistent with state-of-the-art seismic requirements. But much of the construction of the DOE complex is old and predates a modern understanding of earthquake ground motion. Current DOE policies are not clear regarding the standards to which the older facilities should be held for the purposes of seismic safety. The effort to improve the seismic capability of older structures is uneven across the complex. There is little or no communication between facilities concerning common problems. The linkage to the outside world is also highly variable, with some sites actively participating with the professional community and others remaining isolated.*

Among the safety issues that must be considered in the design and operation of weapons facilities is the response to ground shaking that may occur because of earthquakes. Modern building codes in the United States reflect the different earthquake probabilities in different parts of the country. For normal construction this practice is a successful one, producing buildings that, for the most part, perform extremely well during earthquakes. For buildings built before the formal adoption of earthquake zoning considerations, such as many of the DOE production facilities, good engineering practices provide a certain amount of earthquake resistance through general specifications and wind loading requirements. Nevertheless, older structures in which hazardous operations take place need to be carefully examined to assess effects of possible earthquake ground motion.

DOE Practice

Although the DOE facilities have not been subject to the same regulatory environment as the commercial nuclear industry, DOE has in fact followed nuclear industry practices for assuring seismic safety as they have evolved, particularly during the past decade. Thus recent construction by DOE reflects standards and practices that are consistent with developments in the commercial nuclear sector. A comparison with the criteria and analyses at nearby commercial nuclear power plants licensed by the Nuclear Regulatory Commission serves to demonstrate that DOE's recent approach conforms to standard modern practices.

A major problem, recognized by all, is that many of the facilities in the complex are old and predate modern earthquake engineering practices. The performance of these structures must be evaluated in the light of modern understanding of earthquake ground motion. With such an evaluation in hand, needed modifications can be made to strengthen the structures or otherwise improve their performance, and strategies can be developed to minimize risk in the event of failure. Changes in other practices, such as anchoring and shelving, may also be indicated.

Earthquake Criteria

Each of the DOE facilities we visited seemed to have an adequate criterion for its "design basis earthquake." The criteria have resulted from a variety of studies, largely probabilistic, that have been sponsored by DOE over the past decade. In addition, some local DOE contractors have undertaken such studies independently.

The studies depend heavily on subcontractors for technical input to augment the small number of internal staff with earthquake engineering expertise.

The Department's weapons complex spans the entire United States and thus encounters the full range of earthquake possibilities. For example, the Savannah River and Oak Ridge sites are in the stable eastern seaboard region, which is characterized by infrequent earthquakes. Although infrequent, earthquakes in this region can be large, as evidenced by the Charleston, South Carolina, earthquake of 1886. A special problem with earthquakes in the eastern United States is that their association with faults is uncertain, making it impossible to predict with any certainty where they are going to occur. A good deal of research is being done in this area, however, and it was encouraging to see the Savannah River staff actively involved in it. SRS also has an active advisory committee, made up of university researchers, helping with local investigations, and there is a special budget allocation for such research. As a result, SRS appears to be following closely all recent technical developments related to earthquake phenomena in this region.

Rocky Flats, Pantex, Sandia (Albuquerque), and Los Alamos lie in relatively stable regions. Earthquakes are infrequent, and their relation to geologically mapped faults is more predictable, permitting the use of conventional methods for specifying design earthquakes and ground motion.

INEL and Hanford are located in the intermountain west, a region characterized by large earthquakes that occur with a frequency exceeded in the United States only in California. As a result, great attention is paid to earthquake phenomena, and there is an ongoing effort to learn more about the seismic potential of faults in the region. These facilities have established internal programs that use site contractor staff, as well as specialized investigations conducted by outside consulting firms. As in the case of SRS, at both INEL and Hanford outside advisory panels or other mechanisms are in place to ensure contact with the professional community.

Lawrence Livermore National Laboratory is located in a highly active seismic region and has an outstanding internal capability in all facets of the science of seismology and of earthquake engineering.

Upgrading Old Facilities

Oak Ridge provides an example of the process being used for seismic review throughout the complex. The specific criterion for the design basis earthquake at Oak Ridge is, of course, different from those at other facilities. Nonetheless, the analysis is typical of that used elsewhere.

All major process facilities at the Y-12 Plant were built in the 1940s or early 1950s before seismic design of facilities became a requirement (see Figure 4.3). For purposes of evaluation, two ground acceleration criteria were established: 0.08 g for facilities with a remaining life of 25 years and 0.12 g for those with 50

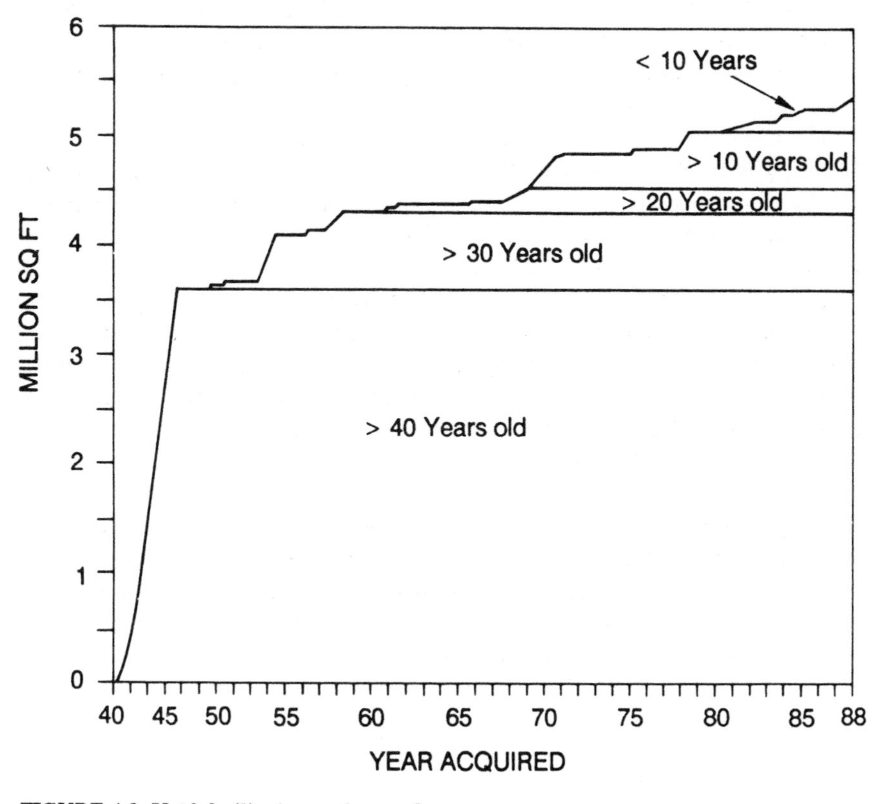

FIGURE 4.3 Y-12 facility base: the age factor.

years remaining. If a new major process facility were to be constructed in Oak Ridge today, it would most likely be designed for something between 0.15 g and 0.18 g. For comparison, the design basis for the Clinch River Breeder Reactor, proposed to be located near Oak Ridge, was 0.25 g. This difference in criteria for plants in the same geologic province, with the same exposure to earthquakes, is not surprising; it reflects the fact that the consequences of an earthquake-induced accident enter into ground-acceleration specifications.

Nonetheless, the use of an acceleration criterion as low as 0.08 g may be questionable. Without a detailed examination of the analysis of the consequences of failure of these older structures, it is not possible to determine whether the approach is sufficiently conservative. Further work is needed to justify this criterion. Given the great range in consequences and the variation in geologic conditions at the different sites, the committee is not in a position to recommend a general priority for seismic upgrading.

The Department and its contractors have focused on seismic threats to buildings.

The current effort is largely performed on a building-by-building basis. Insufficient attention is paid to seismic issues affecting systemic safety over the site as a whole, such as earthquake damage to emergency systems, communications, and fire-fighting capabilities.

There appears to be a commitment from management to provide the resources necessary to identify problems. Where reviews identify minor modifications that can be made to improve earthquake resistance, they are implemented rapidly. However, should major renovation be called for as a result of these reanalyses, it is not clear how priorities would be assigned.

Currently, the emphasis is on safety of operation. That is, the analyses seek to assure that the damage will be sufficiently limited to prevent a major release of radioactive material. But even damage at an "allowed" level could terminate operations indefinitely. In the future it may be necessary to add to the evaluations some cost-benefit considerations concerning the possible loss of production capability in the event of an earthquake.

Recommendation *The Department of Energy should develop improved guidelines for seismic review of older structures housing hazardous facilities. A uniform policy should be established that takes into account realistic estimates of remaining useful life and costs and benefits so that sensible assignment of priorities for seismic upgrading of older structures can be made.*

5
Health

The production of nuclear weapons involves activities and materials that can affect human health adversely. Some of the hazards differ little from those encountered in other industrial settings, and they are addressed by the usual practices of industrial safety and hygiene. Others, such as exposure to radiation and the use of certain hazardous materials, are unique to the weapons complex. Both workers and the public are potentially affected. Occupational health programs are instituted to protect workers, while environmental controls are established to protect public health. The principles of occupational health management apply to both chemical and radiation hazards.

Exposure to radiation can have a variety of effects on human health. Effects range from degenerative illness or death to cancers, developmental abnormalities, and possibly, genetic changes. These effects can occur at both high and lower levels of exposure. The severity of effects generally decreases with decreasing dose. Based on extrapolation from animal studies, the effects of gamma or x-ray radiation decrease as the dose is delivered over a longer period of time.

Studies of atomic bomb survivors and persons exposed to therapeutic or diagnostic radiation provide information on effects on health that result from higher levels of exposure. However, there is not sufficient information to confirm the methods used for extrapolating from these data to the much lower levels of exposure that normally arise in the occupational setting. Studies of animals are limited by the extent to which quantitative extrapolations to humans can be made, and even these studies are not unambiguous.

We are not aware of any widespread occurrences of occupational diseases in the nuclear weapons complex arising from failure to comply with recognized and acceptable work practices. When sporadic occurrences have arisen, the causes were, for the most part, readily identifiable. Some cases of disease caused by accidental exposure to high levels of radiation or to other hazardous substances, such as beryllium, have occurred in the past, but current procedures have been improved to minimize these exposures. However, we believe there is a recognized tendency to reduce vigilance when, over an extended period of time, no obvious or immediate adverse effects arise. We are concerned that there is a notable absence of aggressiveness at most facilities in addressing the possible, although as yet unidentified, effects on health from long-term, low-level occupational exposures.

OCCUPATIONAL HEALTH

Occupational health or occupational medicine, broadly defined, concerns "all aspects of the relationship between work and health," and it "is to a large degree concerned with the impact of work on the development of medical disorders" (NRC 1988a). The discipline of occupational health comprises three interrelated kinds of activities: (1) prevention of illness that could occur because of occupational exposures or hazards, (2) diagnosis and treatment of illness in the work force, and (3) monitoring and surveillance of the exposure of workers.

Each of these activities is pursued in the weapons complex to varying degrees. Health physics, industrial hygiene, safety, and emergency planning are carried out with the aim of preventing illness and injury. Medical departments currently provide clinical therapeutic services and employee assistance programs in the course of diagnosis and treatment. Monitoring and surveillance of workers are carried out under several different auspices, including the health physics, industrial hygiene, and medical departments. The protection of human health is central to each of these efforts.

The Role of Medical Expertise

Conclusion *The central focus of programs in health physics, safety, industrial hygiene, emergency planning, and medical programs is the protection of human health; but occupational medical input to decisions is inadequate.*

In general, the interaction among the health physics, safety, and industrial hygiene professionals seems to be effective. There is, however, negligible occupational medical input to decisions. As best we could determine, the medical departments at the respective weapons facilities become involved with exposures in the workplace only in special cases. For example, at some sites employees working with especially hazardous materials, such as beryllium or lead, are part of a monitoring program that involves the medical department. Similarly, medical attention is provided to employees with injuries or those who have incurred a potentially harmful exposure to radiation or chemicals above established limits. But the medical departments are rarely involved in decisions related to monitoring and controls in the workplace. Often, data on the exposure of employees to radiation or chemicals are stored elsewhere in a facility and become part of an employee's health record only after an inquiry from the medical department. Moreover, medical personnel must often rely on their patients to provide the names of chemicals or other hazards to which they may be exposed in the workplace. In short, medical departments are for the most part relegated to a reactive role.

The role of the medical department in DOE headquarters appears to mirror those in the contractor facilities. At DOE headquarters, the Office of the Medical Director is located in EH and consists of only four employees: a physician who is the director; a Ph.D., who is assistant director; a program manager; and a secretary. The medical department at headquarters is responsible for administering medical programs and auditing contractor programs. But it is not sufficiently powerful to reshape a program so as to assure more timely and effective medical input. Our review of recent audits indicates that the DOE medical department recognizes the deficiencies identified above, but it has no authority to correct them.

Recommendation *The occupational health programs at the weapons complex facilities should be improved by encouraging collaboration among the industrial hygiene, health physics, medical program, and other health-related functions at each facility. In particular, medical expertise already available in the facilities should be integrated into the daily decisionmaking aspects of all occupational health activities. The DOE headquarters medical department should be given sufficient resources to administer, monitor, and effect change in these programs.*

Chemical Hazards

Conclusion *While substantial progress has been made to control exposures to ionizing radiation in the weapons complex, as evidenced by the significant reduction in occupational doses, there appears to be a less than adequate emphasis on hazards associated with exposures to chemicals routinely used in industrial operations.*

Certain potentially hazardous agents, like beryllium and asbestos, are subject to special control efforts. However, weapons production operations require the routine use of many other chemical agents that present potential hazards to workers, such as cutting oils, organic solvents, and plating solutions. Agents suspected or known to be hazardous require one or more of the following types of control: training, employee monitoring, special handling, or removal of the worker from the hazard, as in the case of pregnant workers. The degree to which the concerns arising from the use of chemical agents in the workplace are addressed varies widely among the weapons facilities (see also Chapter 4).

Recommendation *Occupational medical programs within the complex should increase the emphasis placed on protection of workers from chemicals suspected to be hazardous at the acute, subacute, or chronic level of exposure.*

ASSESSING RISKS TO HEALTH

In addition to implementing adequate protection programs, it is essential that health procedures be evaluated to determine their effectiveness and identify areas for improvement.

Monitoring in the Workplace

Conclusion *Collection of health-related data concerning workers in the weapons complex is inadequate with regard to both the kinds of data collected and how they are stored.*

Exposure to high concentrations of hazardous agents characteristically results in manifestations of acute toxicity. Acute exposure also may result in disease after a period of many years. Chronic exposure to chronic toxic agents may result in disease only after the passage of long intervals of time, often even after the cessation of exposure.

Because serious effects on health can arise from chronic exposure to hazardous agents at low levels, worker populations should be tracked using a multidisciplinary monitoring and surveillance program. Programs of this type are an integral component of an effective occupational health program; they include periodic medical examinations, industrial hygiene and/or health physics measurements in the workplace, bioassays to measure exposures, and other studies as appropriate. The findings from ongoing monitoring and surveillance programs provide data from which adverse effects resulting from low-level occupational exposures can be identified at the earliest possible time.

Such data about workers are useful only to the extent that they are accurate, comprehensive, accessible, and comparable among different populations of workers. Current technology can be used to enter information about the exposure of workers into their medical records automatically. The type of data collected should be standardized within the weapons complex, and it should extend beyond radiation exposures to include toxicants, carcinogens, and reproductive toxicants. Data should be stored so as to allow comparisons not only within a given facility but also among facilities. Finally, the data should be accessible to analysis by researchers.

The medical department at DOE headquarters has recognized the need for such an effort and has recently retained an outside contractor to develop a computerized data management package called the Health Track System. The system is intended to provide the capabilities described above, and as designed it has promise. It is not clear, however, how this system will be coordinated with the Secretary's recently proposed Comprehensive Epidemiologic Data Repository (CEDR) (see below). Further, its success will depend on the ability of the DOE headquarters medical department to effect the necessary changes among DOE contractors.

Recommendation *Monitoring and surveillance programs in the complex should be improved substantially through the use of standardized protocols for data collection, storage, and analysis. In particular, data collected within the complex should be comprehensive, accessible, and comparable.*

Research on Effects of Exposure to Low Levels of Radiation

As a result of the nuclear power plant accidents at Three Mile Island and Chernobyl, public concern regarding the adverse effects of ionizing radiation on human health increased. Included in this increased concern was the health of workers within the weapons complex and that of people living in its vicinity. The Department's record of vagueness and secrecy regarding releases of radioactivity and the extent of environmental contamination contributed significantly to a public lack of confidence in DOE's concern about risks to human health.

It has been known since the early 1900s that ionizing radiation is carcinogenic to humans. Current understanding of the health effects of radiation is based on data obtained from studies of survivors of the atomic bombings of Hiroshima and Nagasaki, and of persons receiving radiation for medical purposes. The radiation doses to these populations are relatively high compared to doses due to background radiation[1] or normal occupational exposure. Estimates of cancer risk at low levels of exposure are obtained from extrapolation from high doses to low doses and depend importantly on the methods used.

The latest estimates of risks from external ionizing radiation are given in BEIR V (NRC 1990) (the effects of internal emitters are covered in BEIR IV [NRC 1988c]). BEIR V provides estimates of low-dose cancer mortality risks based on extrapolation of cancer rates induced from high doses of ionizing radiation. From these values one can compute the risks to hypothetical, highly exposed workers (approximately 10 mSv/yr) or to people exposed to low doses (approximately 0.01 mSv/yr), as shown in Table 5.1. The BEIR V estimates lead to the conclusion that it is barely possible to detect the carcinogenic effects of 10 mSv/yr on an adult

[1] Background radiation, excluding radon and medical exposures, is variable but roughly near 1 mSv/yr (National Council on Radiation Protection and Measurements 1987). The current DOE radiation standards for protection of the public (DOE Order 5480.1A) are 1 mSv/yr for prolonged exposure and 5 mSv/yr for occasional annual exposures for a maximally exposed individual.

TABLE 5.1 Average Lifetime Cancer Mortality per 100,000 Males (Extrapolated from Effects of Acute Exposure to Radiation)

Dose (mSv/yr)	Exposure Period (yr)	Leukemia		Nonleukemia[a]	
		Background[b]	Excess[c]	Background[b]	Excess[c]
10	18-65	780	400	20,140	2,480
0.01	1-99	790	0.7	19,760	4.5

[a]A dose rate reduction factor has not been applied to the risk estimate for nonleukemia cancers. Suggested values for such a factor range from 2 to 10 for low dose rate exposures NRC 1990.
[b]PHS 1984.
[c]NRC 1990.

population of 10,000 and is impossible to detect the effects of 0.01 mSv/yr on 100,000 people.

For comparison, the calculated dose to the "maximally exposed individual" (a hypothetical nearby resident considered to receive the highest possible radiation dose from the facility) at Hanford in 1987 was 0.0005 mSv/yr (Pacific Northwest Laboratory 1988). Based on the estimates from BEIR V, health effects on the public resulting from this level of exposure from the Hanford site should be negligible.

Nonetheless, near the two British fuel reprocessing operations, Sellafield and Dounreay, clusters of excess cases of leukemia among children and young adults have been reported in a population that is believed to have received only low doses (Darby and Doll 1987, Forman et al. 1987). Although no general increase in other cancers among other age groups has been associated with living in the vicinity of nuclear installations in England and Wales, there is some suspicion of increased multiple myeloma and Hodgkin's disease in adults that requires further study (Forman et al. 1987, Cook-Mozaffari et al. 1987). Clusters of cancers may be found in population-based epidemiologic studies for a variety of reasons. These include: (1) the estimates in Table 5.1 are grossly inaccurate; (2) the doses have been grossly underestimated; (3) statistical artifacts; or (4) the unexposed, control population to which the study population is compared may not have been appropriately matched. For example, it is possible that there may be a viral cause of childhood leukemia in situations of large-scale population mixing of the sort that takes place in the construction and operation of new nuclear facilities (Kinlen 1988). Therefore comparisons to leukemia rates in more stable populations would be inappropriate in this case.

The resolution of the uncertainties in the interpretation of cancer clusters requires more data and can only be done from large epidemiologic surveys involving many millions of person-years of exposure. Several international epidemiology studies on low-level radiation and cancer are under way at the International Agency for Research on Cancer (IARC). All European cancer registries are being used to investigate the possible association of childhood leukemias with exposure to radioactive material from the accident at Chernobyl in 1986. In 1988 an international group including DOE representatives began the design of an analysis of radiation worker cohorts and their risk of cancer (preliminarily, a dose-related increase in risk of multiple myeloma was observed). It is estimated that data involving 1.7 million person-years of exposures will be available for this effort (J. Esteve, personal communication, IARC 1989). Designs are also under way to study international utility reactor worker populations with a total of approximately 2 million person-years of exposure (Esteve, personal communication, September 29, 1989).

Although the weight of current scientific evidence suggests that, at the estimated levels of exposure, leukemia clusters reported in the British studies are not the result of radiation exposure, the committee believes that a similar study should be conducted in the vicinity of a DOE facility. Such a study should investigate at least childhood leukemia and adult multiple myeloma. The study could provide relevant information in a different setting and may also serve to assure the local population.

Epidemiology on Exposure of Workers

Conclusion *The Department has compiled a substantial body of data on the exposure of workers to radiation and the status of their health, and it is seeking means to ensure that epidemiologic studies will be conducted using these data.*

Since the early days of operations in the nuclear weapons complex, the exposure of workers to radiation has been monitored by detection devices, called badges or dosimeters, that are worn by the worker throughout his or her work shift. The data are often stored on incompatible data bases separate from other information about the health of workers, an arrangement that makes it difficult to access and analyze the data.

In the recent past, DOE has conducted studies on several groups of workers at selected sites in an effort to relate radiation exposure to cancer mortality. The studies have been conducted primarily by researchers at the DOE national laboratories, although some of the work was carried out at universities under contract to DOE. More recently, the mortality of workers at Oak Ridge has been investigated (ORAU 1988). These research efforts, in spite of their obviously crucial importance, have been disappointingly limited thus far. Studies have focused almost exclusively on gamma-radiation exposures and cancer endpoints.

Greater consideration should be given to other radiations, as well as exposures to chemicals, and other effects on health.

The Secretary of Energy has recently adopted a four-point initiative on epidemiology as a first step toward addressing these concerns. The initiative calls for the establishment of the Secretarial Panel for the Evaluation of Epidemiologic Research Activities (SPEERA) to provide the epidemiology program with guidance on policy issues, such as goals, management and reporting structure, resource requirements, quality control, records maintenance and access, and similar concerns. The National Research Council's Committee on the Radiation Epidemiologic Research Program (CRERP) is charged to guide the program on scientific issues. Moreover, the Department is committed to the development of a Comprehensive Epidemiologic Data Repository (CEDR), which is to contain relevant data on all current and former DOE contract employees. (These employees number approximately 600,000.) The data will be made available to independent researchers when CEDR becomes operational. The Secretary has also mounted an effort that entails assistance from CRERP to allow independent researchers access to these data before CEDR becomes operational.

The Department should be encouraged in its efforts to seek outside advice in areas related to research on epidemiology and dose reconstruction. The involvement of external committees like SPEERA and the National Research Council's CRERP will not only strengthen the technical basis of these efforts, but will also lend credibility to the findings. We believe it is essential that the research be conducted by independent researchers who are trained in epidemiology.

Recommendation *The Department of Energy should compile data on its workers in a comprehensive, comparable, and accessible data base and should support epidemiologic research using these data. Furthermore, the DOE worker health research studies should be designed and directed independently and subject to peer review by an external organization.*

Epidemiology on Exposure of the Nonworker Population

Conclusion *In spite of the limitations imposed by low levels of exposure and small population sizes, dose reconstruction and epidemiology studies can provide a useful mechanism for addressing public concerns about the potential for adverse health effects in areas near weapons facilities.*

The Department of Energy recognizes that it must address public concerns arising from past and current exposures to radiation released from the weapons complex facilities. This is a singularly difficult task because the interpretation of the scientific data and the assessment of risk are complex, and public perceptions may not be in accord with scientific understanding. Carefully designed, independently directed dose reconstruction and epidemiologic studies could advance scientific knowledge and may help to improve public understanding.

The conduct of such studies requires a careful determination of both exposures and effects. Since exposures to nonworker populations are not measured, doses must be reconstructed by calculations based on whatever data are available. Exposures in recent years can be quantified by using data collected at monitoring stations in communities located near these facilities. In the absence of such monitoring data, dose reconstruction studies combine data on releases of radioactive materials with calculations of the transport and fate of those materials (i.e., how they might move through the environment and ultimately result in an exposure) to estimate levels of exposure to the population. Dose reconstruction studies are, of course, limited by the quality of the available data on releases and by the accuracy of the transport models. It is important that the results of these studies include a thorough analysis of the uncertainties involved.

Several dose reconstruction studies are now being conducted in the United States. For example, the National Cancer Institute is supporting work at the University of Utah to conduct dose reconstruction and epidemiologic studies of populations living in the vicinity of atmospheric tests (Wachholz, in press). At Fernald, a dose reconstruction study is being conducted by the Centers for Disease Control. DOE is also supporting a dose reconstruction study at Hanford to estimate the exposures of residents near the facility as a result of historical releases (Till, in press). The Hanford investigation involves the reconstruction of exposures in the 1940s and 1950s as well as the design and execution of a thyroid cancer epidemiologic study by the Centers for Disease Control. The National Cancer Institute is also currently conducting an epidemiologic study of cancer mortality in the vicinity of commercial nuclear power plants in the United States (Jablon et al. 1988).

The use of independent agencies, panels, and peer-reviewers serves to strengthen the scientific validity of research efforts and to increase the credibility of results. At Hanford, for example, the use of an independent agency and review panel is greatly enhancing both the scientific quality of the study and public confidence in it.

Each DOE facility is engaged in unique operations that result in the use of differing combinations of radionuclides and chemicals and thus present unique sets of exposures. As a result, attention to potential health effects should be tailored to each facility.

The combination of a dose reconstruction study and epidemiologic analysis can be used to ascertain whether there is evidence of an association between exposure and a consistent pattern of disease. Statistical analyses are performed to determine the probability that findings result from an actual effect rather than simply by chance. Thus in studies where exposure levels are relatively small and the increased incidence of disease is slight, the study population must be large enough to confer statistical significance on the results. Nonetheless, such studies should be undertaken to improve scientific understanding and inform the public.

Recommendation *The Department of Energy should continue to support dose reconstruction and epidemiologic studies of relationships, if any, between exposure to low levels of radiation from the facilities and the incidence of disease to improve scientific understanding and to inform the public. The studies should be designed and directed independently in a manner that involves public participation and external peer review.*

6
Modernization of the Complex

Most of the physical plant of the nuclear weapons complex is, in a word, old; many of the processes employed, generally dating from the 1940s and 1950s, are old-fashioned. Consequently, opportunities—and challenges—exist not only for refurbishing the plant but also for introducing alternative processes that could improve overall efficiency and facilitate the attainment of health, safety, environmental, and production goals.

In this chapter, we review selected modernization issues and describe some technological opportunities for improving current and future operations, including remediation of existing waste sites. We recognize, however, that modernization of plants and methods costs money. Decisions to decommission existing buildings, to build new production facilities, to rebuild existing ones or to take advantage of new technologies must take account of the benefits to be gained for the costs incurred, including opportunity costs.

THE DOE MODERNIZATION REPORT

At the request of Congress, DOE (1988) prepared a report on the modernization of the complex, projecting its configuration to the year 2010. Congress asked that the study consider "... the overall size, productive capacity, technology base, and investment strategy necessary to support long-term national security objectives." The DOE study emphasizes that the mission of the complex is to supply the DOD stockpile requirements and, at the same time, to maintain technological superiority and comply with health, safety, and environmental requirements. The Department also considers flexibility in the production capability of the complex to be another important requirement.

The Department's report, while not necessarily adopted as guidance by the current administration, is the only available planning document with a long-term focus. The report recommends changes for the complex in three categories, in order of priority (see Figure 6.1). The first category includes activities considered essential that must be accomplished in the near term. In this category are remediation of existing inactive waste sites, as well as compliance with applicable regulations at currently active waste disposal sites, refurbishment of plutonium recovery capacity in Building 371 at the Rocky Flats Plant, construction of new production reactors and processing facilities at the SRS and INEL, and construction of an SIS facility at INEL for the enrichment of fuel-grade plutonium into weapons-grade plutonium.

The second category—deemed essential but not urgent—includes upgrading facilities for processing virgin plutonium at SRS; upgrading the Y-12 Plant facilities for processing uranium; upgrading, renovating, and modernizing facilities and laboratories throughout the complex; and establishing facilities at SRS, INEL, and the Hanford Nuclear Reservation for vitrifying mixed hazardous and radioactive wastes for eventual permanent storage.

The third priority includes objectives considered optimal for the future, although their phasing would have to depend on the availability of funds. This category includes permanently closing the Feed Materials Production Center (FMPC) at Fernald; eliminating the weapons programs at Hanford; relocating the activities currently performed at Rocky Flats; and relocating the materials operations at the Mound Facility. Without specifically commenting on each of the proposed changes in the renovation and modernization report, we focus on two broad issues: the capacity for processing plutonium and the need for maintenance.

Capacity for Processing Plutonium

Most of the activities of the complex focus on the production, separation, and preparation of plutonium and tritium. Obviously, the expected future demand for the production of these materials thus must be evaluated in order to guide the modernization of the complex. Indeed, because the projection of demand provides the foundation for long-term planning, it is important that DOE and the Congress obtain the best and most objective advice that is available on this point.

We have no special information or expertise that enables us to assess the current or future requirements that are or might be imposed by the President's Stockpile Memorandum. Nonetheless, some general observations can be made. Given a level of demand for new or refurbished weapons, the production capacity for tritium is the more problematic because tritium is a highly perishable isotope (i.e., it has a short half-life, 12.3 years). The situation is different regarding

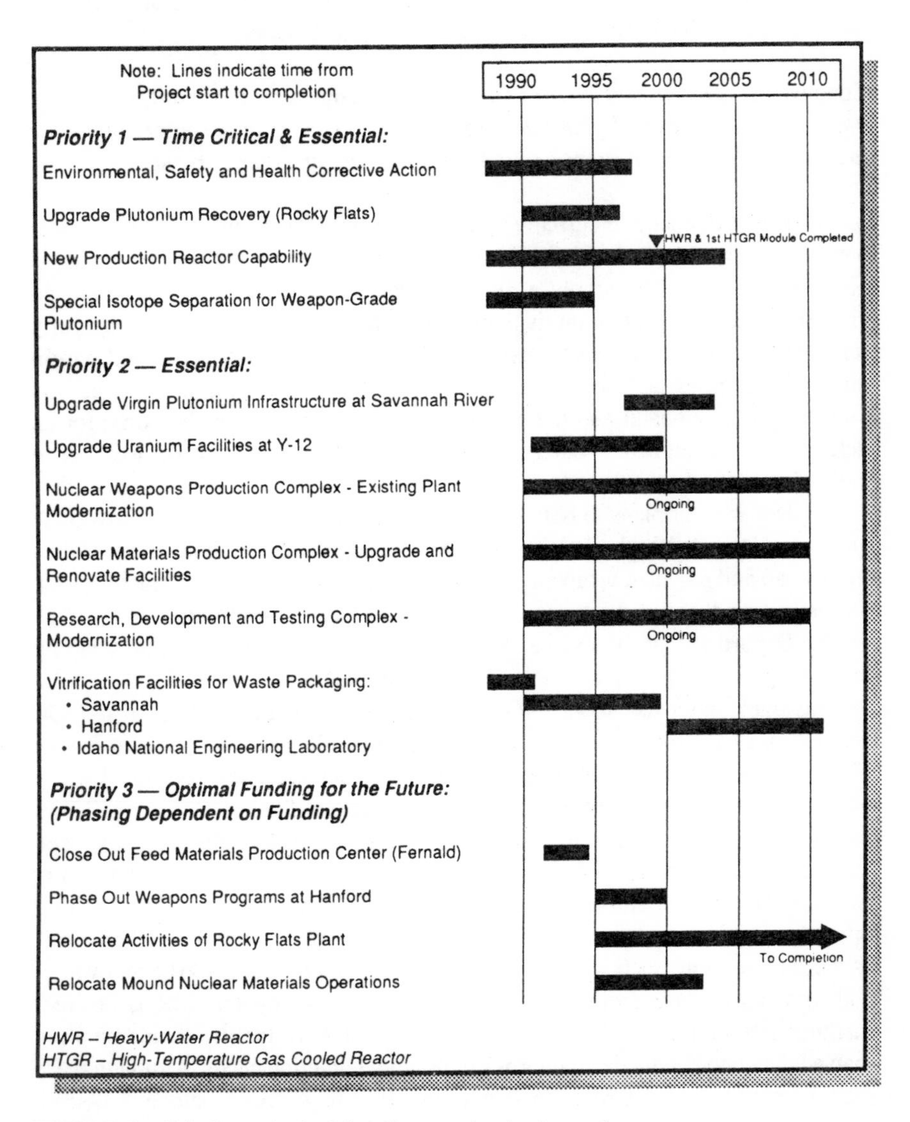

FIGURE 6.1 Priority and schedule of key modernization actions.

plutonium; its current supply in the stockpile of weapons, scrap, and spent reactor fuel is large and its half-life is very long, about 24,000 years.

Conclusion *The current supply of plutonium and the current capacity to process both virgin and recycled plutonium from retired weapons or scrap are adequate to meet the demand for maintaining a stockpile similar to the current one.*

The national stockpile currently contains several tens of thousands of nuclear weapons. The plutonium in these devices, plus that in the supply chain, is obviously sufficient to supply a nuclear deterrent of the existing size or even greater. Because plutonium is long-lived and toxic and must be carefully safeguarded for reasons of national security, the production of additional, virgin plutonium implies additional costs to society for maintaining safeguards and protecting public health and the environment. These costs must be borne for an indefinite time, and hence, other things being equal, it is not sensible to produce more plutonium than we need.

The Department plans to obtain additional capacity to process weapons-grade plutonium by using both chemical and isotope separation methods to recover it from scrap and recycled weapons and by laser isotope separation of reactor-grade plutonium produced in the N-Reactor at Hanford (see Appendix B).

The Department proposes to add to its capacity to process plutonium scrap by renovating Building 371 at Rocky Flats at an estimated cost of $400 million. Serious questions exist about the cost-effectiveness of this renovation if DOE concludes, as the modernization report urges, that all operations now at Rocky Flats should be moved elsewhere. Moreover, the need for additional scrap recovery capacity is doubtful. The $90 million New Special Recovery (NSR) facility, also designed for plutonium scrap processing, is already in an advanced stage of construction at SRS. And the Plutonium Facility (Building TA-55) at LANL is an efficient and productive operation for scrap recovery. This facility, operating for the most part on a one-shift, 5-day schedule, can process almost half as much plutonium as Rocky Flats can (even if Building 371 were to be renovated) and turn out a purer product. If additional capacity beyond NSR is desired, institution of a three- or four-shift operation at the LANL facility should be more than adequate to handle the complex's plutonium recycling needs. Although there may be resistance at LANL to converting Building TA-55 into a full-scale production facility, an administrative solution should be possible. In any case, more extensive use could be made of this efficient operation with its exemplary operating history and its strong technical staff.

The development of isotope separation technology is approaching the pilot plant stage at LLNL. DOE proposes to construct a production-scale SIS facility at INEL at an estimated cost of $600 million. Plutonium containing concentrations of plutonium-240 greater than 7 percent is undesirable for use in weapons (see Appendix E). Plutonium containing more than 7 percent but less than 13 percent plutonium-240 can be converted to weapons-grade material by blending it with plutonium containing much smaller concentrations of plutonium-240 obtained from Savannah River. Thus SIS would be used to process plutonium having more than 13 percent plutonium-240 to obtain purer material to be used in blending.

The weapons complex inventory of reactor-grade plutonium containing more than 13 percent plutonium-240 is located at Hanford and amounts to about 7 or 8 tonnes. But, to our knowledge, no compelling need for this material has been demonstrated, nor are there currently forseen uses for the SIS facility after the reactor-grade plutonium has been processed.

Special isotope separation also introduces important new considerations relating to safety and safeguards. First, SIS is the first process that involves the vaporization of plutonium in a high-vacuum system. We have no reason to believe that this process will create a major new hazard that cannot be managed; but the new technology raises environmental controversies, and considerable effort is required to demonstrate that concerns about human health and the environment can be satisfied. Second, SIS introduces a potentially undesirable precedent with respect to nonproliferation goals (NAS 1985). By introducing technology for converting reactor-grade to weapons-grade plutonium, it forms a potential bridge between the civilian fuel cycle and weapons production. Spent civilian power reactor fuel contains substantial quantities of plutonium, but this fuel contains concentrations of plutonium-240 sufficiently high that, in the absence of SIS, it would be undesirable for use in weapons. Federal law prohibits the use of spent civilian reactor fuel for nuclear explosive purposes (42 U.S.C. 2077). Once developed, the SIS technology could be applied in other countries, including those not now possessing such weapons, greatly increasing the quantity and improving the quality of materials from which nuclear weapons could be built (NAS 1985). Any decision to proceed with the SIS facility should explicitly consider the implications of the technology for nuclear proliferation.

Recommendation *The Department of Energy should concentrate on making better use of the existing plutonium processing capacity as required and postpone plans to construct additional capabilities.*

Renovation and Modernization

The Department's modernization report calls for an annual outlay of 4 percent of the replacement value of the physical plant per year for renovation and modernization. The allocation is evidently based on a rule of thumb that is applied in industry to estimate maintenance expenses. Without clearer understanding of how the renovation and modernization activities envisioned in the report relate to maintenance, we find it difficult to comment on the adequacy or the basis of this allocation.

As discussed in Chapter 2, we found the level of attention paid to maintenance in the past to have been generally inadequate, and we support the improvement of efforts in this area. We also noted that, given the special nature of the complex, common rules of thumb may not apply. Determining the level of funding needed for maintenance and the allocation of resources for the purpose varies with circumstances and should be determined as the result of the usual budgeting process. Although increased maintenance does impose costs, the benefits can include greater safety, as well as improved reliability and availability of equipment.

OPPORTUNITIES FOR ADVANCED TECHNOLOGY

The Department's 5-year plan for environmental remedial action and waste minimization includes a research and development program that is aimed at the demonstration of new technology for these purposes. We strongly concur with the emphasis in this plan on the need for advanced technology in waste management and remediation, and we agree that the research and development necessary to achieve it is vitally important.

The Department and its contractors should also be alert to opportunities from other sources to introduce new technology or to employ more benign materials, thereby improving the effectiveness of the complex in meeting production goals in a way that is consistent with health, safety, and environmental objectives. Over the past decade or two, private industry has increasingly recognized the importance of using technology that meets these multiple objectives, particularly in minimizing the generation of wastes. Developing or taking advantage of advanced technology is an essential ingredient in the success of private industry, and it can be no less valuable in improving the efficiency of the complex.

For example, the complex generally employs costly, old-fashioned metal-forming processes typical of foundries and machine shops, perhaps because these were the only processes available when the complex was originally designed. Unfortunately, foundry and machine-shop processes typically create significant quantities of scrap and substantial problems of waste management. Indeed, a substantial portion of DOE's processing efforts is dedicated to recycling the scrap materials generated in these processes. Moreover, the scrap and waste problems are exacerbated in the case of weapons production by requirements for safeguards and by the hazards of radioactive and toxic materials. Perhaps alternative processes exist that could increase both efficiency and safety in the use of special nuclear materials and, at the same time, minimize problems of maintaining safeguards and managing waste. Perhaps significant long-term savings might in fact be realized by using more modern and efficient processing technologies.

We have not made a comprehensive survey of the technology opportunities that are available to the complex. That task is a daunting one, particularly if undertaken from the outside and from the top down. In the course of our review, however, we considered several particularly important opportunities that can serve as examples.

Upgrading the Chemical Processing of Plutonium

Conclusion *When the weapons complex was originally designed, chemical processes for the separation of plutonium were based on fluoride chemistry. These processes create substantial problems of toxicity, corrosion of equipment, and exposure of workers to radiation hazards. Safer alternative processes are now available.*

Historically, the conversion of plutonium solutions to metal has involved a multistep process based on fluoride chemistry (see Appendix D). The process, which is based on extractive metallurgical procedures in use for many years, has been used to recover plutonium since the days of the Manhattan Project in World War II. It has the advantages of reliability and relative ease of operation. Unfortunately, it also has disadvantages.

Gaseous hydrogen fluoride and aqueous hydrofluoric acid are exceptionally corrosive. Both are also highly toxic and have properties that exacerbate the problem: hydrogen fluoride, being a gas, is readily mobile, and hydrofluoric acid has the ability to penetrate the skin, causing systemic poisoning. Moreover, plutonium fluorides emit copious neutrons from alpha-n reactions—approximately 200 times as much as is emitted from plutonium oxides. Neutron exposure can be reduced by shielding enclosures and equipment, but effective shielding often impedes operations because of its clumsiness. It is better to remove the source of radiation than to try to shield against it.

Viable alternatives to fluoride-based plutonium processing exist. Frequently, for example, plutonium in recycled weapons can be subjected to molten-salt extraction to remove americium (the main contaminant of concern) and then refabricated for reuse. Less pure plutonium requires more extensive chemical processing, but the fluorination step can be bypassed by direct oxide reduction (DOR), in which plutonium oxide (PuO_2) is reduced directly to metal with calcium. Yields from DOR are lower than those from plutonium fluoride (PuF_4) reduction (but in either case the sludge must be reprocessed), and the product may require electrorefining to achieve the desired purity. The reduced yields, however, may be offset by the lower costs associated with reduced hazards and lower maintenance requirements so that the net result may be a the lower total cost per unit of plutonium produced.

Even if not all the steps in its multistep production are replaced, the total use of fluoride processing can be reduced. Specifically, replacement of the fluorination steps in existing peroxide and oxalate precipitation-based processes appears to offer net advantages.

Such a drastic process modification would take time and money to introduce at some facilities in the complex, and alternative processes may require additional development effort before they can be made suitable for full-scale application to production. Nevertheless, there is little doubt that these processes can become effective replacements for the existing fluoride-based technology. The capability already exists at LANL.

Recommendation *As it proceeds in its modernization efforts, DOE should give priority to replacing any needed capacity for plutonium conversion processing that currently is based on fluoride chemistry with technology based on safer, less corrosive materials that may offer lower total costs when proper maintenance, health, safety, and environmental factors are taken into account.*

Computing and Communications Technology

Conclusion *The Department of Energy nuclear weapons complex can make better use of computing and communication technologies to improve performance, particularly in operational areas like training, safety, process control, and management.*

Within the weapons complex, computing and communication technologies are actively used in a diversity of applications, although such use is inconsistent and less than optimal when viewed across the complex as a whole. Some of the best expertise in scientific computing in the world resides in the laboratories. Notable examples of success imported from outside sources exist at the facilities in obvious areas such as accounting, management, inventory control, and documentation. Successes are less visible in operational areas such as process control, training, and event or status logging. The potential for application of computer technology spans virtually all aspects and levels of operations across the complex and constitutes an opportunity for significant, sustained improvement in performance and safety.

The world's base of computing technology continues to grow, driven by advances in very large scale integration, data storage and systems, and most significantly, accessible computers, networking, and application software. While the DOE laboratories are among the leaders in scientific computing, which they pioneered for studies in such areas as reaction physics, thermomechanical behavior, and scientific data analysis, the production facilities lag behind the state of the art in applying computing tools to field operations.

Opportunities for broad exploitation include the following.

- *Simulators for training operators.* Training resources and techniques vary widely across the complex: most of the installations rely on classroom training and operations manuals. At SRS, training of chemical process operators, as well as reactor operators, incorporates computers, simulators, and full-scale replicas. These advanced techniques are extremely effective in giving operators detailed understanding of processes and operations and should be adopted across the complex.

• *Operational monitoring of tank transfers.* The use of computers has significantly reduced errors that occur in this common operation at some facilities, but the use is not widespread in the complex.

• *Event logging.* Records ranging from shift activities to logging of field status reports are now prepared mechanically for the most part. Such logging could be extensively computerized with obvious benefits for identifying outliers immediately, reconstructing events, establishing trends, and other tasks.

• *Schedules and planning.* Using computers to optimize factory processing would allow flexibility in production practices in a modernized complex. The Y-12 Plant has apparently been applying computers for such tasks successfully.

• *Medical data.* Collection of data on the health of workers and records of exposure using automated data systems that are available commercially would facilitate data access and analysis.

Some groups within the complex have in fact been developing software applications to improve the performance of the weapons complex, and we envision that computing will inevitably play an ever more critical role in its safe and environmentally sound operation.

Recommendation *The Department of Energy should encourage and facilitate computer use as it affects operations, health, safety, and the environment throughout the complex. The Department should promote local and complexwide networking to archive and disseminate successful practices. Specifically, DOE should develop and apply computing technologies of critical and specific relevance to the weapons complex, such as training simulators, process controllers, and event loggers.*

Robotics

Conclusion *The Department of Energy can make better use of robotics and remote technology in performing the work of the weapons complex.*

Robots refer here to electronically controlled mechanisms that perform useful work. The weapons complex has special needs for robotic devices of many types. They include mobile work systems of the kind used at Chernobyl and Three Mile Island, stationary devices that service hot cells and package waste, automated excavators that can exhume buried waste, material-handling robots for repositories, and automated machining and processing robots of the kind appropriate for the modernized complex of the year 2010.

The application of robots within the complex should depend on the nature of the task, risk, robotic competence, and cost. While the most universal motivation behind the use of robots is the escalating cost of manual operations, another impetus is the effort to cope with conditions that are threatening to humans, such as acute exposure to radiation during emergencies, exposure to contamination in waste-handling operations, and activities in constricted work spaces. In such circumstances, robots can have great advantages over manual alternatives. Robots are also obviously useful for repetitive tasks that demand high precision but that workers may find boring.

The application of robotics or even an awareness of the robotics state of the art varies significantly throughout DOE. Most of the sites have at least fledgling programs in robotics or experience with components that could become the building blocks for more complicated applications. But overall, the weapons complex has generally not taken advantage of more recent advances in robotics. Although the earliest remote manipulators were pioneered for nuclear hot-cell work, subsequent technological evolution was driven more by advances in subsea activities and by missions of the military, the manufacturing community, and most recently, the space program.

Numerous opportunities exist now for applying robotics throughout the complex, but certain targets emerge at specific sites. Of course, successful demonstrations anywhere can always be made more broadly applicable. Examples of opportunities include the following.

- *Emergency response.* To our knowledge, the complex does not have a viable fast-response force with expertise, devices, personnel, and transportation at the ready in the event of emergencies that limit human response. The responses at Three Mile Island and Chernobyl were hampered by just such a lack of remote equipment, and they focused the world's attention on the need for it.
- *Buried tanks (single- and double-walled).* Aged, faulty, and contaminated tanks are a generic problem throughout the complex. Robots could play a significant role here in inspection, remedial action, and as necessary, decommissioning. Constricted spaces like the annulus of double-walled tanks also preclude human entry and call for the use of robots.
- *Excavation.* Buried wastes, such as those in trenches at the Y-12 Plant, are candidates for unmanned excavation, but the most visible, voluminous, and imminent application is the exhumation of acres of transuranic and mixed wastes at INEL. Robotics is clearly the technology of choice in such applications.

Other opportunities include inspection; characterization and cleanup of ductwork; subsurface mapping, particularly prior to excavation; maintenance of hot cells and repositories without human entry; facility decontamination and decommissioning; and unmanned production processing.

Robotics has the potential to reduce costs and risks significantly, but cost projections must be examined with care: the use of robots involves large up-front investments in engineering and equipment. Opportunities may exist for DOE to incorporate existing robotic capabilities developed for other applications, but in certain cases, the conditions under which robots might work in the complex may place special requirements on the systems. Examples affecting design include the need for radiation-tolerant components and consideration of decontamination for servicing or replacement.

Recommendation The Department of Energy should expand its use of robotics technologies wherever they can be applied to fulfilling the critical and specific needs of the mission of the weapons complex cost effectively.

Appendixes

Appendix A
Biographical Sketches of Committee Members

RICHARD A. MESERVE is a partner in the Washington law firm of Covington & Burling. He holds both a law degree from Harvard Law School and a Ph.D. degree in applied physics from Stanford University, where he did postdoctoral work on the theoretical properties of paramagnets and techniques to calculate molecular properties. In 1976, he was a clerk for Supreme Court Justice Harry A. Blackmun, and in 1977 he was appointed Legal Counsel and Senior Policy Analyst in the White House Office of Science and Technology Policy (OSTP). At OSTP he helped develop policies designed to promote the technological advance of American industry and conducted reviews of energy technology issues. In addition, he served as executive director of an interagency committee concerned with nuclear power plant safety. Mr. Meserve has been a member of several study committees of the National Research Council, including the Panel to Study the Impact of National Security Controls on International Technology Transfer. Recently, he served as chairman of the National Research Council Committee to Assess Safety and Technical Issues at DOE Reactors.

RONALD L. ATKINS is head of the Chemistry Division of the Research Department of the Naval Weapons Center at China Lake, California. His division conducts research and development across a broad spectrum of scientific disciplines focused on materials science. He received a Ph.D. degree in organic chemistry from the University of New Hampshire. Dr. Atkins' research expertise is in the synthesis and characterization of highly energetic materials for explosive, propellant, and pyrotechnique applications. He also has conducted research in high-density materials synthesis, high-energy laser dye synthesis, and alternative synthetic fuel development.

ALBERT CARNESALE is professor of public policy and academic dean, John F. Kennedy School of Government, Harvard University. His expertise is in nuclear energy policy and nuclear weapons policy. He has worked as a scientist with the U.S. Arms Control and Disarmament Agency and served as an adviser to the U.S. delegation to the Strategic Arms Limitation Talks. He was head of the U.S. delegation to the International Nuclear Fuel Cycle Evaluation and served on the National Research Council's Board of Radioactive Waste Management. He is a member of the American Nuclear Society. He holds a Ph.D. degree in nuclear engineering from North Carolina State University.

JESS M. CLEVELAND is chief of the Transuranium Research Project of the U.S. Geological Survey. His current research focuses on the environmental chemistry of the transuranium elements, especially plutonium. He has been employed both at Rocky Flats and Hanford Laboratories (now Pacific Northwest Laboratories); in the course of 35 years of experience in plutonium chemistry, he has investigated its analysis, processing, fundamental chemical properties and environmental behavior and is author or co-author of over 30 scientific papers on the subject. He is author of the book *The Chemistry of Plutonium* and coauthor of *The Plutonium Handbook*. He was a member of the National Research Council's Transplutonium Working Group and has served on the program review committee of the Earth Sciences Division of Lawrence Berkeley National Laboratory. He holds a Ph.D. degree in inorganic chemistry from the University of Colorado.

DAVID G. HOEL is the director of the Division of Biometry and Risk Assessment, National Institute of Environmental Health Sciences, National Institutes of Health. He received his Ph.D. degree in statistics from the University of North Carolina. His research is focused on the quantification of human health risks using molecular, toxicological, and epidemiological data, with particular interests in radiation. He has served as an associate director at the Radiation Effects Research Foundation (RERF) in Hiroshima, and is currently a member of the BEIR V Committee of the National Research Council on ionizing radiation. Dr. Hoel has been active as a member of international and U.S. government advisory committees including the Office of Science Technology Policy's Committee on Cancer Policy. He is currently a member of the Council of Fellows of the Collegium Ramazzini and is a member of the Institute of Medicine of the National Academy of Sciences.

GEORGE M. HORNBERGER is a professor in the Department of Environmental Sciences, University of Virginia. His research focuses on hydrological processes, particularly as they affect the transport of solutes and particulates through soils and rocks. He serves on the Board of Radioactive Waste Management of the National Research Council. He received a Ph.D. degree in hydrology from Stanford University.

PAUL KOTIN is currently a consultant on pathology and adjunct professor of pathology, University of Colorado Medical School. Formerly, he was senior vice president for health, safety, and environment at the Johns-Manville Corporation. Previous to that he served at Temple University as dean of the school of medicine and vice president for health sciences. He is also a former director of the National Institute of Environmental Health Sciences and scientific director for Etiology, National Cancer Institute. His work focuses on the mechanisms of carcinogenesis and on environmental factors relating to cancer. He is the recipient of the Knudsen Award from the American Occupational Medicine Association, was the Gehrmann Lecturer for the American Academy of Occupational Medicine, and was recipient of the Distinguished Service Medal of the Secretary of Health, Education, and Welfare. He is a fellow of the College of American Pathologists and a member of the American Association of Cancer Researchers and the American Association of Pathologists and Bacteriologists.

DENNIS J. KUBICKI is a fire protection engineer in the Office of Nuclear Reactor Regulation of the Nuclear Regulatory Commission. He is responsible for reviewing fire protection programs at nuclear power plants and determining their degree of conformance with Nuclear Regulatory Commission guidelines and requirements. Previously, he was assistant manager for Industrial and Fire Safety in the Safety Office of the National Aeronautics and Space Administration. He has extensive experience in evaluating fire protection for buildings, residences, institutions, and municipalities. He is a member of the National Fire Protection Association, the Society of Fire Protection Engineers, and the American Society of Safety Engineers. He holds an M.S. degree in safety from the University of Southern California (Eastern Division) and a master's degree in business administration from the University of Maryland.

J. CARSON MARK has retired from his position as leader of the Theoretical Division of Los Alamos National Laboratory. His research is involved with group theory, transport theory, hydrodynamics, and neutron physics. He has served on the Science Advisory Board of the U.S. Air Force, and was a member of the Nuclear Regulatory Commission Advisory Committee on Reactor Safeguards. He is a member of the American Physical Society and the American Mathematical Society.

MICHAEL R. OVERCASH is a professor in the Department of Chemical Engineering and also in the Biological and Agricultural Engineering Department at North Carolina State University. He is also center director of the large

Research Center for Waste Minimization and Management of the U.S. Environmental Protection Agency. His work is focused on fundamental process changes to reduce waste generation and chemical loss and the evaluation and disposal of toxic wastes, including site remediation. He manages a significant research group in terrestrial effects of industrial organic compounds, including greenhouse studies on plant response and soil fate of chemicals. He has served on many scientific and governmental panels on these and related topics. He was recently a member of the North Carolina Governor's Waste Management Board and has drafted guidelines for sludge land treatment in Delaware. His testimony before the congressional Committee of Science and Technology explored the reasons why technologies available for eliminating or treating hazardous wastes have not had an impact on landfill usage in the United States. In 1986 he was awarded the Environmental Protection Agency Distinguished Scientist Award. He is a member of the American Institute of Engineers and the Institution of Chemical Engineers, London.

WOLFGANG K.H. PANOFSKY is currently director and professor emeritus at the Stanford Linear Accelerator Center (SLAC). He was director of SLAC from 1961 to 1984 and prior to that was director of the High Energy Physics Laboratory at Stanford University. He has served with many working groups and commissions, including the President's Science Advisory Committee and the International Union of Pure and Applied Physics. He has received the E.O. Lawrence Award (1961), the California Scientist of the Year Award (1967), the National Medal of Science (1969), and the Franklin Institute Award (1970) and is also the recipient of the Enrico Fermi Award (1979). He is a member of the National Academy of Sciences and is a fellow of the American Physical Society.

RICHARD L. SAGER, JR., is loss prevention manager at the Lithium Corporation of America in Gastonia, North Carolina. He is responsible for overall guidance of the Corporate Safety Program, including staffing and overseeing the Health Services Department, ensuring that all four company facilities are in compliance with corporate and federal safety regulations, and coordinating the Industrial Hygiene Program to establish baselines. Previously, he was supervisor for safety and environmental engineering for several mining and ore processing companies. He has extensive experience with the regulations of both the Occupational Safety and Health Administration and the Mining Safety and Health Administration. He holds a B.S. degree in mechanical engineering from Vanderbilt University. He is a member of the American Society of Safety Engineers and serves on the Executive Committee of the Metals Section of the National Safety Council.

RICHARD B. SETLOW is associate director for life sciences and was former chairman of the Biology Department at Brookhaven National Laboratory. His work in molecular biophysics has studied the effect of ultraviolet and ionizing radiations on proteins, viruses, and cells. His research includes studies of nucleic acids and their repair mechanisms, as well as studies on environmental carcinogenesis. He received the Finsen Medal in 1980 and the Enrico Fermi Award in 1988. He is a member of the National Academy of Sciences and the American Academy of Arts and Sciences. He received a Ph.D. in physics from Yale University.

DAVID R. SMITH has recently retired from his position as leader of the Criticality Safety Group in the Health Division at Los Alamos National Laboratory. He has served as consultant on criticality reviews at many facilities, including Argonne, Oak Ridge, and Hanford. He has a long involvement with developing and maintaining national and international safety standards for the handling, processing, and storing of fissile material, working with the American Nuclear Agency Technical Committee on Criticality Alarm Systems. He is a fellow of the American Nuclear Society.

STEWART W. SMITH is currently on leave from the University of Washington to serve as president of the Incorporated Research Institutions for Seismology, a 58-member consortium of U.S. universities that is developing new global and portable seismographic networks with the support of the U.S. National Science Foundation. At the University of Washington he served as professor and chairman of geophysics from 1970 to 1980, and is presently professor of geophysics and adjunct professor of geological sciences. His research areas include seismicity, tectonics, and earthquake hazards. He has published widely in these fields and served on numerous scientific and governmental panels concerning these topics. One such recent assignment was as chairman of a workshop for the congressional Office of Technology Assessment concerning the seismic verification of nuclear testing treaties. He has had extensive experience in the siting of commercial nuclear power plants and other important structures in earthquake-prone regions.

JOHN E. TILL is president of Radiological Assessments Corporation, located in Neeses, S.C. He is also president and owner of the Embeford Dairy Farm. His expertise is in the assessment of radionuclide releases to the environment. Major fields of interest are the development and implementation of accurate and reliable environmental monitoring programs and the consolidation of current radiological assessment methods to provide a concise evaluation of models and their ranges of applicability. He has received the Elda E. Anderson Award of the Health Physics Society and has served on the Executive Committee Science Advisory Board of the U.S. Environmental Protection Agency. He is a consultant to the Nuclear Regulatory Commission Committee on Reactor Safeguards and a member of the National Cancer Institute Committee on Assessment of Dose due to Fallout Radioiodine. He holds a Ph.D. degree in nuclear engineering from the Georgia Institute of Technology.

VICTORIA J. TSCHINKEL is senior consultant with the law firm of Landers, Parsons, and Uhlfelder in Tallahassee, Fla. Previously, she was secretary of the Florida Department of Environmental Regulation. She was instrumental in rewriting Florida's water quality standards and in revising the state's groundwater protection rules. She has served on the DOE Energy Research Advisory Board and on an U.S. Environmental Protection Agency Panel on toxic pollution and is currently a member of the DOE Advisory Committee on Nuclear Facility Safety. She received an A.B. degree in zoology from the University of California, Berkeley.

F. WARD WHICKER is a professor in the Department of Radiology and Radiation Biology, Colorado State University. His areas of expertise include the transport of radionuclides through aquatic and terrestrial ecosystems and the effects of radiation on plant and animal communities. He has been active in the development of computer simulation models for pathway analysis and dose assessment. He spent one recent sabbatical year as adjunct senior ecologist at the Savannah River Ecology Laboratory. He has served as consultant and committee member for many industrial and governmental studies. He is a fellow of the American Association for the Advancement of Science and a member of the Health Physics Society and the Ecological Society of America.

WILLIAM L. WHITTAKER is director and senior research scientist of the Field Robotics Center of the Robotics Institute at Carnegie-Mellon University. His research focuses on the development of robots for many diverse applications, including reconnaissance and cleanup in nuclear environments, hazardous waste management, planetary exploration, automated mining, autonomous haulage, and automated excavation of buried pipes. His work on teleoperator transporters for radiological emergency response and recovery found direct use in remote systems for recovery of the reactor basement at Three Mile Island. For his robotics work he has received the Science Digest Top 100 U.S. Innovators Award. He holds a Ph.D. degree in civil engineering from Carnegie-Mellon University.

GEROLD YONAS is president of Titan Technologies, located in San Diego. Previously he was Chief Scientist and Acting Deputy Director of the Strategic Defense Initiative Organization (SDIO). Before joining SDIO he was Director of Pulsed Power Sciences at Sandia National Laboratories. His work at Sandia was involved with intense electron and ion beams, high power microwaves, lasers, high density plasma, radiation sources, and energy conversion devices. Dr. Yonas is a recipient of the Secretary of Defense Medal for Outstanding Public Service. He is a fellow of the American Physical Society, a member of the American Institute of Aeronautics and Astronautics, and has served on the executive Board of the controlled Nuclear Fusion Division of the American Nuclear Society. Dr. Yonas resigned from the committee in May 1989 when he returned to managerial responsibilities within Sandia, which was one of the facilities covered by this study.

Appendix B
The DOE Nuclear Weapons Complex: A Descriptive Overview

Nuclear weapons are produced in the United States by the Office of Defense Programs (DP) of the Department of Energy (DOE). DP manages a large complex of facilities, including 17 major plants in 12 states, to carry out its mission. Its annual budget is about $10 billion, and DOE and its contractors employ about 80,000 people.

For the purpose of description, we organize the facilities in the complex into three main types: the weapons laboratories, the materials production facilities, and the weapons production facilities. The weapons labs design the weapons—providing the blueprints and technical specifications for their construction—and test them. The materials production facilities provide the raw nuclear materials for fabrication into warheads. The weapons production facilities fabricate the required nuclear components, supply the hundreds of non-nuclear components, and assemble the warheads. In addition, DP manages the test facility and a waste repository for the operation, which is currently the Waste Isolation Pilot Plant in New Mexico.

Figure B.1 presents a snapshot of the current status of the complex, which, over its 40-year history, has been configured in many different ways. It would be impractical to indicate all the previous routings. It can be seen that the complex has redundant capabilities for many processes, and in some cases, processes once required for stockpiles have been discontinued.

WEAPONS LABORATORIES

The three weapons laboratories are the Los Alamos National Laboratory (LANL) in Los Alamos, New Mexico; the Lawrence Livermore National

Laboratory (LLNL) in Livermore, California; and the Sandia National Laboratory (SNL) in Albuquerque, New Mexico.

Los Alamos National Laboratory and LLNL are both large multipurpose complexes, and they both conduct many activities unrelated to nuclear weapons. Within the context of nuclear weapons production, however, these two labs have essentially the same missions: to design, develop, and test the nuclear components of the weapons. Both labs are operated under contract with the University of California Board of Regents. Between the two labs there is a vigorous competition, which has undoubtedly been beneficial to weapons design. Each lab has provided designs currently in the weapons stockpile, and each lab has several new designs under development.

Sandia National Laboratory, like the other two labs, carries out many activities not associated with nuclear weapons. Part of its mission in the nuclear weapons complex is to design and engineer non-nuclear components associated with a nuclear weapon. Such components include electrical systems, fusing and firing, neutron generators, tritium reservoirs, and delivery packages. SNL works closely with LANL and LLNL to incorporate the nuclear components of a new design into an operational weapon. SNL also has responsibility for engineering modifications and upgrades to weapons already deployed and for monitoring the stockpile.

MATERIALS PRODUCTION FACILITIES

The materials production facilities include the gaseous diffusion plants at Oak Ridge in Tennessee, Paducah in Kentucky, and Piketon in Ohio; Fernald, Ashtabula, Hanford; the Idaho Chemical Processing Plant, Y-12 at Oak Ridge, and the Savannah River Site.

The mission of these facilities is to provide the nuclear materials used in nuclear weapons, particularly uranium-235, uranium-238, plutonium-239, lithium-6, tritium, and deuterium. The first three are among the heaviest of elements, while the latter three are among the lightest. Four of the six are produced by separation from naturally occurring ores and water. The other two, plutonium and tritium, are not available from natural sources but are produced in nuclear reactors by transmutation of other elements.

Heavy Metal Production

Uranium as found in the ground is mostly uranium-238 with 0.7 percent uranium-235 and 0.01 percent uranium-234. Uranium mills process the ores to produce a concentrated uranium oxide, U_3O_8, which then is commercially converted to gaseous uranium hexafluoride (UF_6) for enrichment processing in the gaseous diffusion plants (GDPs). There are three GDPs: the Oak Ridge GDP (K-25 Plant) in Tennessee, the Paducah GDP in Kentucky, and the Portsmouth GDP in Piketon, Ohio. The Oak Ridge GDP has been on standby since 1985. The purpose of these

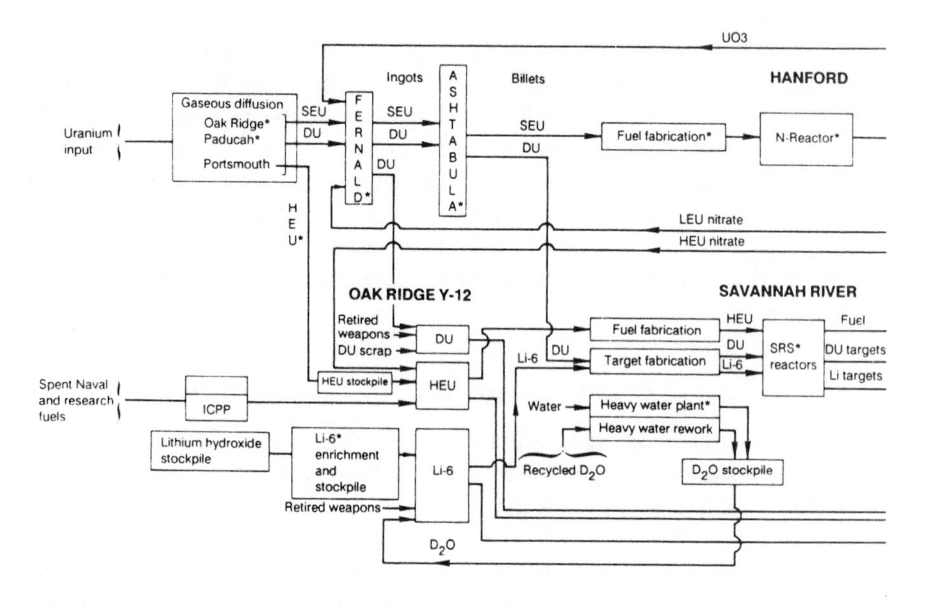

*Not in operation at present

FIGURE B.1 Flow of materials through the DOE Nuclear Weapons Complex. Diagram by P. Rapp 1989.

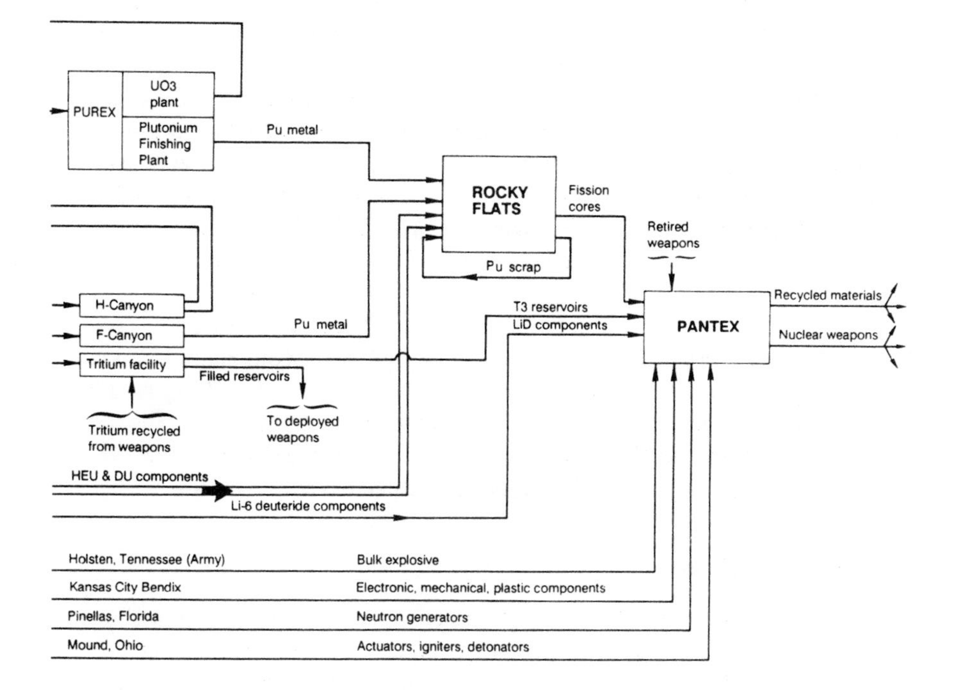
PUREX
UO3 plant
Plutonium Finishing Plant
Pu metal
ROCKY FLATS
Fission cores
Retired weapons
Pu scrap
H-Canyon
F-Canyon
Tritium facility
Pu metal
T3 reservoirs
LiD components
PANTEX
Recycled materials
Nuclear weapons
Filled reservoirs
Tritium recycled from weapons
To deployed weapons
HEU & DU components
Li-6 deuteride components
Holsten, Tennessee (Army)
Bulk explosive
Kansas City Bendix
Electronic, mechanical, plastic components
Pinellas, Florida
Neutron generators
Mound, Ohio
Actuators, igniters, detonators

plants is to concentrate the uranium-235. The mechanism for concentration is based on the fact that a UF_6 molecule containing uranium-235 is slightly lighter than a UF_6 molecule containing uranium-238; consequently, the former has a slightly higher thermal velocity. About two-thirds of the uranium-235 in the natural ore is removed in the concentration process, so that there are two product streams: enriched uranium and depleted uranium. Only the Portsmouth GDP is now operated to provide concentrations higher than 4 percent uranium-235.

In general the enriched uranium may contain any percentage of uranium-235. Uranium containing more than 20 percent uranium-235 is called highly enriched uranium (HEU); enriched uranium with less than 20 percent uranium-235 is known as low enriched uranium (LEU). HEU production for weapons ceased in 1964. Before that time, the gaseous HEU was shipped from Piketon to the Y-12 Plant, where it was converted to metal and stockpiled. The HEU metal is commonly known as "oralloy," where the first two letters indicate Oak Ridge. LEU was used for the fuel/target rods at the Hanford N-Reactor, which is now on cold standby. For purposes of comparison, the fuel in commercial power reactors is about 3 percent uranium-235, the driver fuel for the production reactors at SRS is typically 60 percent uranium-235, and naval reactor fuel is 97.3 percent uranium-235. Depleted uranium (DU) is used both for SRS target rods and for components in weapons.

There are two, almost independent, plutonium production streams. Both streams start with enriched and depleted uranium in the gaseous state, and both streams provide plutonium metal to Rocky Flats. By far the largest effort in heavy metal production is devoted to the creation and processing of plutonium. The first stream to be described here flows through Fernald, Ashtabula, and Hanford. The second goes through INEL, Oak Ridge, and Savannah River.

Fernald/Ashtabula/Hanford

The LEU and DU products, still in the gaseous state, are shipped to the Feed Materials Production Center (FMPC) in Fernald, Ohio. At the Fernald plant the UF_6 is reduced to the "green salt," UF_4, and mixed with green salt produced from other inputs to the FMPC. The FMPC is a large and diverse facility containing 10 separate plants. Uranium input to the FMPC enters in several forms, including ore concentrates, metal scraps and residues, uranyl nitrate (UNH) from SRS, and UO_3 from Hanford. All these inputs are processed into uranium oxides, some of which may be shipped to the GDPs for enrichment. Most of the input, however, is processed into UO_3 and hydrofluorinated into green salt, which is mixed with the green salt from reduction of UF_6. The green salt is reduced to metal and cast into ingots. Some of the DU is shipped to Y-12 for fabrication into weapons components. The LEU ingots, and the rest of the DU ingots, are machined and shipped to the Ashtabula Extrusion Plant in Ashtabula, Ohio. In Ashtabula the ingots are extruded into tubes and billets for later fabrication into reactor rods. The DU tubes are returned, first, to Fernald for further machining, and then to

SRS where they are used as target rods for transmutation of uranium-238 into plutonium. The LEU billets go to Hanford to make fuel/target rods for the N-Reactor, which is currently on cold standby. The DOE Modernization Report anticipates that the FMPC will be permanently closed in the near future.

At the Fuel Fabrication Facility on the Hanford site, the LEU billets are fabricated into reactor elements by extrusion into rods clad with zirconium. These rods serve as both fuel and target in the N-Reactor. Neutrons from fissioning uranium-235 convert some of the uranium-238 to various isotopes of plutonium. The fissioning isotope, plutonium-239, is the one desired for both reactor fuel and weapons. As the uranium-235 is used up, the amount of plutonium increases, but the fraction of plutonium as plutonium-239 decreases as the relative abundance of plutonium-240 increases. "Weapons-grade" plutonium contains less than 7 percent plutonium-240, while "fuel-grade" plutonium contains less than 13 percent plutonium-240. Other isotopes, plutonium-241 and plutonium-242, are also produced by subsequent neutron capture. Consequently, the limit to the N-Reactor fuel cycle is determined not by the burn-up of uranium-235, but rather by the desired abundance of isotopes in the produced plutonium. Chemical processing of the irradiated reactor rods separates plutonium from the other elements. Chemical processing cannot, however, separate plutonium isotopes. Methods to do that are still under development.

Chemical processing of the irradiated rods begins in the PUREX (plutonium-uranium extraction) Plant on the Hanford site. The first step is the chemical removal of the fuel cladding in the head-end dissolver. Subsequently, the fuel is initially dissolved in an aqueous solution of nitric acid. An organic solvent is used to separate the nitrates of uranium, plutonium, and neptunium from the fission products. Further treatments with organic solvents and nitric acid solutions isolate the uranium, plutonium, and neptunium. The three major outputs from the PUREX Plant are UNH, plutonium oxide (PuO_2), and neptunium nitrate. The neptunium is shipped to SRS; the other two products are processed further at Hanford.

The uranyl nitrate goes to the Uranium Oxide (UO_3) Plant, where it is calcined into uranium oxide powder. The powder is shipped either to the GDPs for enrichment or to the Fernald Plant for processing into metal.

The plutonium oxide goes to the Plutonium Finishing Plant (PFP, or Z Plant). There the plutonium is precipitated as the oxalate, converted to PuO_2 and fluorinated with gaseous hydrogen fluoride (HF). The resulting pink powder, PuF_4, is reduced to metal with calcium (see Appendix D). The plutonium is shipped to Rocky Flats for fabrication into weapons components. PFP can also be used to recycle scrap plutonium from Hanford and Rocky Flats.

ICPP/Y-12/Savannah River

The other heavy metal stream is somewhat newer, and it depends to some extent on heavy elements previously produced. An important input to this stream is the spent naval fuel returned to the Idaho Chemical Processing Plant (ICPP), located on the site of the Idaho National Engineering Laboratory (INEL). Other inputs to ICPP include spent fuels from research and test reactors, both domestic and foreign. The main mission of ICPP is the recovery of highly enriched uranium for use in driver fuel in the SRS reactors. ICPP has extensive water-filled storage and staging facilities, allowing fuel to be moved into head-end dissolution without exposure to air. The head-end facilities offer a variety of dissolution processes to accommodate the various fuels and claddings. Subsequent processing is similar to the PUREX process, involving solvent extraction and purification. ICPP is distinguished by its capability for handling and recovering the highly enriched uranium of the returned naval fuels. Naval fuel returns are expected to increase rapidly, perhaps tripling over the next 10 years. The output product, powdered uranium oxide, is shipped to the Y-12 Plant at Oak Ridge for processing into SRS fuel. A secondary mission of ICPP is the recovery of krypton-85 from the spent fuels. The krypton-85 is shipped to Oak Ridge for commercial sale, largely for use in the detection of leaks. ICPP is the only source of krypton-85 outside the Soviet Union.

The Y-12 Plant at Oak Ridge is a large multi-purpose facility with several different missions, both in materials production and in weapons production. One mission is to produce uranium metal of about 60 percent enrichment for use as SRS driver fuel. Because the several inputs to this metal production have varying enrichments, the process streams are carefully blended to produce the required enrichment. One input is the ICPP oxide, which may have originated with naval fuel or with reactor fuels. The other input is highly enriched uranyl nitrate from processing of spent fuels at SRS. To blend the product to the required enrichment, oralloy from the Y-12 stockpile is added to the mix, and the final uranium metal product is shipped to Savannah River.

Unlike the Hanford N-Reactor, which uses reactor rods functioning simultaneously as fuel and target, the SRS reactors use independent fuel and target rods, made of different materials. At the SRS Fuel Fabrication Facility, the uranium metal from Y-12 is alloyed with aluminum and extruded into fuel rods with aluminum cladding. At the SRS Target Fabrication Facility, the hollow tubes of depleted uranium from Fernald are electroplated with nickel and bonded into aluminum cans to serve as target rods for transmutation into plutonium. The Target Fabrication Facility also assembles the lithium target rods for tritium production (see Light Element Production below).

There are five production reactors at SRS, designated C, K, L, P, R. They were all designed with heavy water, D_2O, as coolant and moderator, allowing great flexibility in the use of the reactors for production of various nuclear materials, including tritium and plutonium. The P, K, and L reactors are currently shut down because of safety concerns; the C-Reactor is being cannibalized, and the R-Reactor is permanently closed. Currently there is no plutonium production planned, at least in the near term, at SRS; the focus is on tritium production.

After the rods are removed from a reactor, they are processed in different chemical separations facilities called "canyons," because they are 30 feet wide by 60 feet high, and almost 900 feet long, heavily shielded in concrete and steel. Too radioactive for human occupation, the canyons are operated entirely by remote control. The PUREX Plant at Hanford is a similar canyon facility. At SRS, the H-canyon processes the discharged fuel rods to recover the enriched uranium, while the F-canyon processes the irradiated DU target rods to recover the plutonium. Tritium is recovered from the lithium target rods in a separate Tritium Facility, where remote handling is not required (see Light Element Production below).

The H-canyon uses a modified PUREX process of dissolution and solvent extraction to recover the enriched uranium from the fuel rods. Highly enriched UNH is shipped to the Y-12 Plant to be recycled and blended back into new SRS driver fuel. LEU, also in the form of UNH, is shipped to the Fernald Plant.

The F-canyon facility also uses the PUREX process, in this case to recover plutonium-239 from the DU target rods. The plutonium nitrate from the canyon is converted to metal by trifluoride precipitation and reduction. The processing to metal is similar to that at the Hanford Z Plant, but it differs in that aqueous hydrofluoric acid is used instead of gaseous HF. The plutonium metal buttons are shipped to Rocky Flats for fabrication into weapons components. A byproduct of plutonium recovery is the recovery of depleted uranium; thousands of drums of DU oxide are now in long-term storage. Recently, the F-area has added a new facility, scheduled to be operational in the near future, for the processing of plutonium scrap.

Light Element Production: Y-12 and Savannah River

The three light elements used in nuclear weapons are deuterium (D), tritium (T), and lithium-6 (Li6). These elements are essential to the fusion process, as distinct from the fission process initiated by the heavy metal elements. In common parlance, the earliest weapons, with only the fissioning elements, are called "atomic bombs," while the modern weapons, upgraded with light-element fusion capability, are called "hydrogen bombs." D is extracted from natural water at SRS, and Li6 is extracted from natural ores at Y-12. T is produced in nuclear reactors from Li6 targets at SRS. Neutron capture by Li6 produces an alpha particle and T.

D_2O Production at SRS

About 0.015 percent of naturally occurring water is heavy water (D_2O). It is extracted from natural water in staged processes of chemical exchange, distillation, and electrolysis. The Heavy Water Plant at SRS extracted D_2O from the Savannah River for more than 30 years, until it was placed on standby in 1984. Heavy water serves as both coolant and moderator in the SRS reactors, and each reactor contains more than 200 tons of it. To maintain its purity, the D_2O is periodically processed through the SRS Heavy Water Rework Unit and returned to the large SRS stockpile. Supplies for weapons production are shipped from the SRS stockpile to Oak Ridge in the form of liquid heavy water.

Li6 and Li6D Production at Y-12

At the Y-12 Plant the heavy water is processed together with metallic Li6 to produce lithium-6 deuteride (Li6D). The Li6D is formed into weapons components and shipped to Pantex (see Weapons Production Facilities below). Li6, like heavy water, was extracted in large quantities at one time, but it is now drawn from existing stockpiles. Li6 is a stable isotope that makes up about 7.4 percent of natural lithium ores. Lithium itself is one of the most abundant elements. During the 1950s, thousands of tons of lithium hydroxide were purchased for the weapons program. Enrichment of Li6 was the mission of several large plants at Y-12. The basis of the enrichment process is the differential preference of lithium-7 for mercury. With the introduction of mercury into aqueous lithium hydroxide, the lithium-7 will concentrate in the amalgam phase. Enrichment of Li6 required the use of very large amounts of mercury at Y-12. Production of Li6 stopped in 1963, after the accumulation of a large stockpile. Li6 components from retired weapons are returned to Y-12 for recycling.

In addition to its use in Li6D components, Li6 is also used for production of tritium in the SRS reactors. Enriched Li6 from Y-12 is shipped to the SRS Target Fabrication Facility, where it is alloyed with aluminum and canned as target rods for the reactors. Li6 is also used for control rods in the reactor cores, as well as for shielding around the core. The irradiated lithium is processed in the SRS Tritium Facility.

Tritium Production at SRS

The Tritium Facility is a "chemical separation facility," but the irradiated Li rods do not require remote-handling canyons for processing. The mission is T separation, purification, and loading. One input is the irradiated lithium-aluminum target rods. The targets are heated under vacuum, and the liberated gases include hydrogen, D, T, helium-3, and helium-4. Palladium diffusion and cryogenic distillation are used to separate and purify the tritium. Another input to the Tritium Facility is the T recovered from deployed weapons which is contaminated with helium-3. Reservoirs filled with tritium are shipped from SRS to Pantex and to military installations. A new facility at SRS, the Replacement Tritium Facility, is almost complete and should begin to operate in 1990. The new facility is underground, and it uses new hydride technology that greatly reduces the amount of tritium in the filling plumbing, and is expected to reduce greatly the releases of T to the atmosphere. This facility will replace the gas handling and processing that is conducted in the existing tritium facility.

WEAPONS PRODUCTION FACILITIES

Production of Weapons Components

Of the six weapons production facilities—Kansas City, Mound, Pinellas, Y-12, Rocky Flats, Pantex—three are involved only with nonnuclear components. The Kansas City Bendix Plant supplies various electrical, mechanical, and plastic components; the Mound Facility manufactures igniters, detonators, and other small-scale pyrotechnic components; and the Pinellas General Electric Plant produces neutron generators and neutron detectors.

The Y-12 Plant and the Rocky Flats Plant contain specialized machine shops that process raw nuclear materials into the finished components required by the warhead designs. The Y-12 Plant bakes and machines Li6D into ceramic weapons components for shipment to Pantex. It also fabricates uranium components, from both enriched and depleted uranium. These components are shipped to Rocky Flats, where they are assembled, together with plutonium and beryllium components, into "pits," i.e., the shells of fissionable materials inside the high explosive of the weapons. The plutonium and beryllium components are fabricated at the Rocky Flats Plant. Many other metal components, including the stainless steel tritium reservoirs, are fabricated in the extensive metal-working facilities at Rocky Flats.

The plutonium input comes partly from Hanford and SRS, and partly from the retirement and scrap recycling operations at Rocky Flats itself. In line with its mission of pit assembly, Rocky Flats also has the mission of disassembling the pits from retired weapons. The recovered plutonium is chemically processed to remove americium, which is purified and shipped to Oak Ridge. Americium is removed either by molten-salt extraction or by dissolution in nitric acid, followed by ion exchange, peroxide precipitation, fluorination, and calcium reduction to metal. This process is also used for plutonium scrap recovery. Building 371 at Rocky Flats was added to the complex in 1981 to modernize and integrate these operations, but it was closed after a short operational run because of design faults.

Assembly and Disassembly of Weapons

Finally, all the components are brought together for assembly at the Pantex Plant in Amarillo, Texas, and from there the devices are delivered to the Department of Defense. The Pantex Plant itself fabricates the chemical high explosives, which are assembled around the pits fabricated at Rocky Flats. Much of the recent work uses modern insensitive explosives, which come in bulk quantities from the Army facility in Holsten, Tennessee. At Pantex the high explosive (HE) is pressed into rough billets and machined to final shape. The HE components are prepared and assembled in special "bays," made of thick concrete and designed to vent an accidental explosion through the earth-covered roof. The bays are spaced to avoid sympathetic explosions.

The final assembly of nuclear weapons takes place in special assembly cells known as "Gravel Gerties." Components going into the final assembly include the high explosives, the pits from Rocky Flats, the Li6D parts from Y-12, the filled T reservoirs from SRS, and the many nonnuclear components from other facilities. Scheduling and staging the shipment and inventory of these components is an intricate business. The completed warheads are staged onsite at Pantex before shipment to military installations.

As a corollary to its mission of warhead assembly, Pantex is also responsible for disassembling retired weapons. The nuclear components are returned to the plants that produce them for processing and recycling. Pantex is the only facility with the capability to disassemble nuclear weapons. It is therefore the starting point for any maintenance or modification of weapons, except for the replenishment of the tritium reservoirs. Pantex also conducts quality assurance testing on components of deployed weapons.

Appendix C
Nuclear Criticality

DEFINITIONS

Nuclear criticality refers to the precise state of an assembly of fissionable material in which one neutron from each fission event causes one subsequent fission. If less than one additional fission results, the assembly is subcritical, while if more than one new fission is caused by each fission the assembly is supercritical. On the average, about 2.5 neutrons are produced by the fission of a uranium-235 nucleus, and about 3 neutrons come from the fission of a plutonium-239 nucleus. Not all neutrons produced in a fission event interact with other nuclei, however. Almost all the neutrons appear immediately in the fission process (prompt neutrons), but a few, something less than 1 percent, do not. The latter are called delayed neutrons; the delay in their appearance can range from less than a second to almost a minute.

The term *critical* in this context means that the assembly uses all neutrons, including those delayed, to maintain criticality. The delayed fraction, under this condition, permits convenient control because small changes in the reactivity of a system are manifest with times characteristic of the delay periods. If, however, only the prompt neutrons are necessary for criticality, the system does not have this controllability. Such a system is said to have achieved "prompt criticality," and the power output will rise very rapidly.

AN ILLUSTRATION

Let us start our illustration with a bare sphere of highly enriched uranium of mass 25 kg. At a density of 18 kg/l, such a sphere would have a radius of 6.92 cm.

Were 100 neutrons to be distributed in our sphere, 65 would leave it without interacting with any uranium nucleus. The remaining 35 participate in nuclear reactions, with three being captured and producing only a gamma ray photon to carry off excess energy (radiative capture). The other 32 neutrons cause fissions, and with 2.5 neutrons per fission, the second generation consists of 80 neutrons to replace our original 100. These 80 in turn will produce another 64, and these 51.2, then 41, and so on. The reproduction factor—or in the terminology of nuclear engineering, the multiplication factor—is 0.8. If we sum the series 100, 80, 64, 51, 41, . . ., the total is 100/(1 – 0.8), or 500. The total number of neutrons produced by fission for each original neutron is referred to as the neutron multiplication; in this illustration the multiplication is 5.

In this assembly of the fissile material, the fission neutrons, which are born with a velocity near the speed of light, traverse the assembly in less than 10^{-9} s (one billionth of a second), and this is approximately the time for one generation. Thus the neutron chain will die out very rapidly.

If we now add 12 kg of the same material to our uranium sphere, for a total of 37 kg, the radius will increase by 0.97 cm. The neutron leakage will be reduced by this shell, and now only 60 of 100 neutrons will escape interaction. Four of the remaining 40 will undergo radiative capture, and 36 will result in fission. The 36 fissions will produce a neutron chain of first 90 neutrons, then 81, 73, 66, 59, 53, . . ., which sums to 900. The multiplication factor is 0.9, and the multiplication is 10.

One more shell of about the same thickness will increase our assembly mass to 50 kg, with radius 8.72 cm. The probability of neutron leakage is now 55 percent, and 45 neutrons will be involved in nuclear reactions, 5 of which will be radiative capture. The 40 fissions will produce 100 fission neutrons, and these will generate another 100, and so on. The multiplication factor is now unity, the assembly is critical, and the multiplication is infinite.

An additional shell with a thickness of about 0.02 cm would change our sphere from delayed criticality to prompt criticality.

In this illustration we have taken some liberties with precise values in the interest of simplification, but the numbers and results are approximately correct.

FACTORS AFFECTING CRITICALITY

Density

The quantity of fissionable material required for criticality (the critical mass) is strongly dependent on the material density. As the density of a system is reduced, leakage of neutrons is facilitated, and more material is required for criticality. For an unreflected system, the critical mass varies inversely with the square of the density. A two-fold reduction in density results in a four-fold increase in critical mass. Delta-phase plutonium has a density about 8/10 that of alpha-phase plutonium, so its bare critical mass is more than 50 percent larger. This effect is very important for handling oxides and other low density compounds, and for storage arrays where the low average density of material permits the safe accumulation of hundreds or thousands of kilograms of material. Increases in material density are not a concern in ordinary situations, but they play a significant role in the design of nuclear weapons.

Moderation

When fissionable material is in solution, or present as finely divided particles, the presence of a "neutron moderator," such as water or a hydrocarbon, can effect a significant reduction in the amount of fissile material required for criticality. The interaction of neutrons with light nuclei, such as hydrogen, lithium, beryllium, or carbon, reduces the neutron energy after only a few collisions. Slow neutrons interact much more readily with nuclei, and in particular they have a far greater probability of causing fission in uranium-235 or plutonium-239 nuclei. There exists an optimum degree of moderation because if the ratio of hydrogen nuclei to uranium nuclei becomes too large, neutron capture in the hydrogen becomes competitive with fission in the uranium.

Reflection

Fissile material can also be surrounded by other materials that reflect neutrons back into the fissile volume, increasing their opportunity for nuclear interaction. Water, for example, is an effective neutron reflector. While the critical mass for a bare sphere of highly enriched uranium is about 50 kg, and for plutonium-239 about 11 kg, reflection by water reduces the mass required by about half. Six inches of water constitutes an effectively infinite reflector. Some other materials are even more effective reflectors, but for purposes of criticality safety, water reflection is commonly assumed to be appropriate because it is not to be expected that close-fitting reflection by other materials will occur inadvertently. The critical mass of uranium-235 at optimum moderation is about 800 g for a reflected sphere, and for plutonium-239 the corresponding value is about 500 g.

Geometrical Shape

The shape of an assembly of fissile material is also a significant parameter in considering the potential for criticality. As the shape of a quantity of material is changed from a sphere to a slender cylinder, leakage of neutrons without interaction is facilitated. For material of any specified composition there exists a cylinder diameter below which criticality cannot be achieved. As an example, for highly enriched uranium nitrate at any achievable concentration, criticality cannot be achieved in a water-reflected stainless steel or boro-silicate glass cylinder of 6 in.

diameter. Process vessels enjoying such characteristics are often referred to as "favorable geometry" vessels. Most criticality safety people avoid the term "safe geometry" because some other fissile material filling the 6 in. cylinder could constitute a critical mass. The favorable geometry vessel may be favorable for only the one material for which it is intended. Also, it is thought that use of the term "safe" might foster a false sense of security.

Neutron Absorbers

Some materials, cadmium and boron in particular, are effective neutron absorbers. Such neutron absorbers may be used to provide criticality control in vessels of large volume or in locations where many vessels are in close proximity and there is concern about neutron interaction occurring between vessels.

ASSESSING CRITICALITY SAFETY

Knowledge of the many conditions under which criticality can occur is fundamental to effective and economical safety in the processing of fissionable material. A substantial body of data on criticality has been accumulated over the past 45 years, much of it obtained in the critical experiment facilities. The facility at Oak Ridge generally specialized in experiments involving uranium solutions; most experiments with plutonium solutions were conducted at the Hanford facility, and Los Alamos provided much of the data on unmoderated, or fast, systems. Both the Oak Ridge and Hanford facilities for studying criticality are now closed. The only other similar facility for criticality studies in the United States is at Rocky Flats, and its use is dedicated to that facility. The French have a fine critical experiments laboratory at Valduc, with which some limited cooperation has been possible.

Computational techniques are also used to assess criticality safety. The capabilities of the large computers are extremely helpful, particularly for interpolation and limited extrapolation of experimental data. It is, however, difficult to develop confidence in calculations that are not tested against experimental data. While such data are available for the materials that are being used in nuclear systems today, more experiments will be required as future progress involves additional materials. Alternatively, large and frequently uneconomic safety margins must be applied.

Effective and efficient criticality control practices are generally recognized to depend on close cooperation between the criticality safety specialist and the process designer and process engineer. The process supervisor usually has the best feeling for the sort of upset conditions that may occur, and the process designer can provide guidance regarding equipment reliability. The criticality safety specialist, who should be skilled at interpreting critical data and evaluating calculations, is responsible for evaluating the degree of criticality associated with conditions that may arise. The objective is to assure that the entire process will maintain an acceptable margin of safety under normal and all credible abnormal conditions.

CRITICALITY ACCIDENTS IN MATERIALS PROCESSING

Unplanned nuclear criticality events have occurred in the past, sometimes with fatal consequences. In process areas, such events have typically produced radiation that is potentially lethal within a distance of about 10 m. None of the eight process accidents that have been reported resulted in an explosive release of energy. All occurred in solutions of fissionable material. Thus nuclear criticality accidents in the weapons complex have historically had consequences comparable to industrial accidents. Since the mid-1960s criticality accidents have occurred at a frequency lower than those characteristic of industrial threats.

The history of criticality accidents is illuminating. The first process accident occurred at the Oak Ridge Y-12 Plant in 1958. Between that time and the middle of 1964 there were five more, or a total of six in about 6 years. These incidents stimulated increased awareness of criticality safety and brought criticality safety organizations into existence at all major processing facilities. In the following 27 years there have been two more such accidents. The improved record must be attributed at least in part to the effectiveness of the safety organizations, staffed largely with people trained at the critical experiment facilities.

All eight process accidents occurred with materials in solutions: three involved plutonium solutions and five involved uranium solutions. Many were associated with off-normal process activities. Three occurred behind heavy shielding and resulted in only modest radiation exposures, but five were basically unshielded and caused two fatalities and numerous significant exposures. None of these occurrences caused radiation exposures off-site or significant damage to equipment.

Four of the process accidents were terminated in less than a few seconds, whereas the other four persisted for many minutes or hours. For all eight, the energy release associated with the first few seconds was about 1 kWh. One persistent reaction had a total energy output of about 100 times this value, but the rate of energy release was moderate. Future accidents may be expected to have similar characteristics because of features inherent to any critical system. The difficulties associated with obtaining a large energy release from an accumulation of fissionable material are demonstrated by the sophistication required of the weapon designer.

Appendix D
Plutonium

Because both plutonium and uranium are fissionable by slow neutrons and are used in nuclear weapons, there is a tendency to think that they have similar physical and chemical properties, but this is not the case. Both are silvery metals, with freshly exposed surfaces resembling iron or nickel in appearance, and both have densities approximately 50 percent greater than lead. Beyond these similarities, the two elements differ widely in their properties. Plutonium is harder and more brittle than uranium, and has a melting point some 500°C (900°F) lower. Although both are relatively easily oxidized, plutonium is much more reactive chemically, and in air it is readily oxidized to plutonium dioxide, PuO_2, the most common form of plutonium in the environment. Two useful reference works covering the properties and chemistry of plutonium are Cleveland (1976) and Comar et al. (1976).

Uranium is less reactive, but it, too, is oxidized, producing a variety of oxides. In contrast to plutonium, which does not exist in nature to a significant degree, uranium occurs naturally in a number of chemical and mineral forms.

Plutonium dissolves more readily in acids, and once dissolved—particularly in nitric acid—its chemistry is so different from that of uranium that the two elements can be chemically separated from each other. Simply stated, the chemical differences between the two elements in solution result primarily from the differences in electrical charges on their ions. Because their ions behave differently, they may be separated from each other by a process known as solvent extraction.

Plutonium is produced in nuclear reactors by the irradiation of uranium-238 with neutrons emitted in the fission of uranium-235. (Natural uranium contains 99.3 percent uranium-238, 0.7 percent uranium-235.) After discharge from the reactor and storage for several months to allow the short half-life fission products (produced by the fission of uranium-235) to decay, the uranium must be processed to remove the few hundred parts per million of plutonium product. The irradiated uranium is dissolved in concentrated nitric acid. After suitable adjustments, this nitric acid solution, containing uranium, plutonium, and fission products, is contacted with an immiscible organic solution of tributyl phosphate (TBP) in a diluent that is essentially a highly purified kerosene fraction. The uranium and plutonium are extracted into the organic phase, leaving the fission products in the nitric acid solution. After additional extraction to remove residual uranium and plutonium, this solution is sent to waste treatment and storage; it is the primary source of high-level waste in the DOE weapons complex.

The organic solution containing the plutonium and uranium is first contacted with a more dilute nitric acid solution containing a "reducing agent" to decrease the electric charge on the plutonium ions, so that they are extracted, leaving only the uranium in the organic phase. The uranium is then extracted into a very dilute nitric acid solution. The nitric acid solution of plutonium is further purified and concentrated by ion exchange, a process in which the plutonium is selectively sorbed onto beds of organic resin while impurities remain in solution and pass through the bed. The plutonium is then removed from the resin (eluted) with dilute nitric acid.

This solvent extraction procedure is known as the PUREX process, and it is used with minor modifications at Hanford, SRS, and INEL. The process can achieve separation factors of uranium from plutonium of greater than 10 million, and of plutonium from uranium of 1 million. Decontamination of fission products from plutonium exceeds 100 million. Recovery of both plutonium and uranium is about 99.9 percent. An additional advantage of the PUREX process is the solid waste minimization: because the primary chemical used in the process is nitric acid, a volatile liquid, it can be removed by evaporation, leaving only a small volume of solid waste. The organic solution of TBP in kerosene, after a simple cleanup step, can be reused.

Preparation of plutonium metal from the nitric acid solution is accomplished by one of several conversion processes which are based on similar chemistry. All three involve precipitation reactions and all require the use of hydrogen fluoride (HF), either as a gas or in aqueous solution. Plutonium is precipitated from the nitric acid solution as the oxalate, peroxide, or trifluoride (the latter only at SRS, using an aqueous solution of HF). After drying, the oxalate or peroxide is converted to PuO_2 by heating in a stream of air. [The trifluoride is converted into a mixture of PuO_2 and plutonium tetrafluoride (PuF_4) by heating in air.] The PuO_2 is then heated in a stream of gaseous HF to convert it to PuF_4, which can be reduced to plutonium metal by reaction with metallic calcium in a pressure vessel. (The PuF_4-PuO_2 mixture produced from the trifluoride precipitate is reduced directly to metal without reaction with gaseous HF.) Reduction yields average 97 to 98 percent for PuF_4 and about 95 percent for the PuF_4-PuO_2. The calcium fluoride reduction slags are dissolved and reprocessed to recover the residual plutonium.

Plutonium metal scrap, calcium fluoride reduction slags, reduction crucibles, and plutonium-containing incinerator ash are dissolved in concentrated nitric acid and purified usually by ion exchange. The purified solution is then treated by one of the conversion processes described above to produce the metal.

The choice of PuF_4 as the plutonium compound for reduction is based on several favorable factors. The large amount of heat released in the reaction of PuF_4 with calcium, combined with the relatively low melting point of the resulting calcium fluoride slag, results in a low viscosity medium that allows plutonium aggregation and thus enhanced yield. In addition, PuF_4, unlike plutonium trichloride, another possible reduction candidate, does not absorb appreciable moisture from the air. (Reduction of compounds with a high moisture content results in excessive PuO_2 formation and lower yield of metallic plutonium.) The principal disadvantages of using PuF_4 are the high neutron fluxes it produces as a result of alpha reactions with fluoride ions, the corrosiveness and toxicity of the aqueous or gaseous HF used to produce it, and the need to use an aluminum salt (typically the nitrate) in dissolving the calcium fluoride slag, thus increasing the volume of solid wastes.

Alternative conversion processes have been studied with varying degrees of success. The nitric acid solution of plutonium may be evaporated and the solid plutonium nitrate converted directly to PuO_2 by heating in air. This procedure, known as direct denitration, is not promising: it tends to produce gummy residues, and the product PuO_2 is inert toward either reaction with HF or direct reduction with calcium. It appears likely that the existing processes involving precipitation and calcination to produce PuO_2 as an intermediate will be retained for the foreseeable future.

It is in the subsequent treatment of PuO_2 that viable alternatives exist. Calcium can reduce PuO_2 directly to metal, but there are problems because the heat evolved is lower than for PuF_4 reduction and the calcium oxide has a higher melting point. The slag is not melted by the heat of reaction, and as a result finely dispersed metal is produced. This problem has been overcome, however, by the use of a molten calcium chloride flux to dissolve the calcium oxide slag and allow the product plutonium to coalesce. The process has found production application at LANL.

Impurities can sometimes be removed from plutonium metal without resort to aqueous processing. Often americium—the impurity of most concern—can be removed from plutonium in recycled weapons by molten-salt extraction using, for example, a sodium chloride-potassium chloride salt containing a few percent plutonium trichloride: the americium, being more reactive, goes into the salt phase and is replaced in the metal phase by more plutonium. Impure metal also may be purified by molten-salt electrorefining procedures using similar salt mixtures. Judicious use of these nonaqueous procedures can, in many cases, simplify processes and increase efficiency, and safety.

Because plutonium reacts with the air with the evolution of heat and because it is a poor conductor of heat, it can be pyrophoric, that is, it can spontaneously ignite in air, particularly when in the form of lathe turnings, which have relatively high surface area and poor contact between individual turnings. Such conditions can promote the build-up of a "hot spot" in a small area that can exceed the ignition temperature of the metal. Several serious fires in the weapons complex have started in this manner. To prevent their recurrence, current practice calls for handling potentially ignitable plutonium in enclosures with a low-oxygen atmosphere.

Since plutonium reacts so readily with the air, it is rarely, if ever, found in the metallic form in the environment. Thus the properties of PuO_2, the common environmental form, are most relevant when attempting to assess the behavior of plutonium. Plutonium dioxide can vary in color from tan to olive green to black, depending on purity and conditions of formation; it should be noted, however, that it is not observed in the environment in quantities anywhere near large enough for its color to be perceived by the eye. Typically, when it is present in soils, for example, it is in the form of a relatively small number of microscopic particles. The density of PuO_2 is high compared to that of most chemical compounds, but only slightly more than half that of the metal.

Nevertheless, individual particles, depending on how they were formed, can vary considerably in density and in aerodynamic properties. Particles are frequently very small and can be subject to short-range atmospheric dispersion under suitable climatic conditions. The dispersion will be spatially nonuniform, but even a small isolated particle can emit appreciable radiation. These factors combine to cause high variability in soil contamination analyses: whether a given soil sample contains high radioactivity or no detectable activity whatever may depend on whether it contains a single "hot particle."

Plutonium dioxide is normally quite insoluble in water and in body fluids (with a few exceptions as noted below); it is even less soluble when formed at high temperature, as in a fire. Hence its dispersion in soil is primarily by mechanical means. It can also be blown along the surface by the wind ("saltation"). It can be washed downward into the soil column by natural factors, and it can be spread both horizontally and vertically by plants and animals. Some limited dissolution of PuO_2 can occur in ocean water and in ground-waters with chemical compositions that enhance plutonium solubility, but this does not generally occur in domestic groundwaters because of their low chemical contents.

The low solubility of PuO_2 in body fluids has several ramifications. Uptake through the gastrointestinal system is small, since PuO_2 is poorly absorbed through the intestinal walls. The most serious modes of entry are inhalation and the contamination of wounds. Once in the body, plutonium can be difficult to remove. Inhaled PuO_2 can be lodged in the lungs for considerable periods of time, and ultimately it works its way into the lymph nodes. Plutonium entering the blood stream through a contaminated wound ultimately deposits in the liver or the bone marrow: in the latter site it can be especially harmful to the blood-forming process. Some success has been achieved in the removal of plutonium from body systems by the use of chemicals known as chelating agents that can dissolve it and allow it to be excreted from the body. Such treatments are more effective when administered soon after contamination, before the plutonium has been "fixed" in the body.

The comparable uranium compound, UO_2 is similar in density to PuO_2, but it is considerably more soluble. Because the common forms—uranium-238 and uranium-235—are much less radioactive than plutonium, the radiotoxicity of uranium is lower. In fact, the primary hazard of uranium ingestion—it tends to concentrate in the kidneys—is chemical ("heavy metal poisoning") rather than radiological.

Appendix E
Physics of Nuclear Weapons Design

FISSION WEAPONS

To obtain a pure fission explosion, it is necessary to have assembled a highly supercritical mass—a mass of fissile material several times larger than the critical mass, considering the particular reflector and density that may apply (see Appendix C). The fissile material and reflector should be in the metal form, so that the neutrons are not slowed down and the chain reaction can build up an extremely high energy density before the forces driving a disassembly can take effect. The reaction will be stopped because of the drop in the density and consequent drop in the criticality (see Appendix C) of the fissile material as it explodes away.

Before it is fired, the fissile material in the weapon must be in a subcritical configuration, and it will require some time to move the material to make the highly supercritical assembly. The final part of this time—after the material has first become critical, but before it has reached the desired fully assembled state—is the "supercritical time," during which the fissile material is capable of sustaining a chain reaction. Because of the extreme speed with which such a reaction can build up, and because of the fact that once the fissile material is vaporized—which sets in when the energy density is only about half a kilogram of high explosive equivalent per kilogram of fissile material—the further progress of the assembly would be halted by the pressures being developed. The result would be that the total energy then generated would be smaller than that intended by a large factor—10, or a 100, or more. The chance of experiencing such a "predetonation" is just the chance that a background neutron may initiate a chain in the time interval through which the system is supercritical during assembly. This depends on both the duration of the interval and the strength of the background neutron source.

Neutron Sources

Cosmic rays provide a universal source of neutrons, but the number of these is smaller than other sources that will be present, being only one neutron per cm^2 every few minutes. This is at sea level: at an altitude of 20 km this source would be about 10^3 times larger.

The dominant sources are those originating from the weapons materials—uranium and plutonium. Each of these is radioactive by alpha decay, and each undergoes some spontaneous fission. The source of neutrons from spontaneous fission is inherent in the material and dependent on the isotopic composition, and nothing can be done about it once the material is chosen.

The rate at which alpha particles are released is also inherent in the material and depends on the isotopic composition; but the rate at which neutrons are produced by the alpha particles is directly proportional to the amount of light element (lithium through fluorine, with the exception of nitrogen) impurities in the material. Efforts to reduce the light element content by chemical purification would be worthwhile at least to the extent that the neutrons produced by alphas colliding with impurities (the alpha-n reaction) not exceed the neutrons produced by spontaneous fission. The boron content may serve as a gauge in this connection because, although it may not be possible to reduce the mass fractions of carbon or oxygen to as low a level as boron, the neutron production from carbon or oxygen at a given mass fraction is a few hundred times smaller than for boron. Also, although beryllium would produce about three times as many neutrons as boron per unit mass, its mass fraction can be reduced by an order of magnitude or more below that of boron.

In highly enriched uranium the neutron source from spontaneous fission is close to 2 neutrons/kg-s, as is the source from alpha-n reactions in such material having 10 parts per million (ppm) boron content. It requires care to reduce the boron content to this level, but it is not extremely difficult to reduce it to a level a few times smaller. Because the rates for alpha decay of the plutonium isotopes are very much higher than for uranium, the alpha-n source in weapons-grade plutonium is about a 1, 000 times larger than that in highly enriched uranium at the same purification level. The difference does not matter, however, because the neutron source from spontaneous fission in weapons-grade plutonium is larger than that in highly enriched uranium by a factor of more than 10,000. The chemical processing of fissile materials in the weapons complex is directed at obtaining very high purity fissile material, and some of the specifications originated in considerations outlined above concerning the alpha-n neutron source.

Assembly Methods

With respect to obtaining a rapid assembly, which is to say a short supercritical interval during assembly, two approaches were used in 1945. One was the "gun-assembly" method in which a subcritical piece of active material (called the projectile) was fired down a gun barrel to mate with a subcritical piece of material (called the target), so that when the projectile was seated in the target the resulting configuration would constitute several critical masses. With a feasible projectile velocity of about 1,000 feet per second (fps), the time interval from first criticality to final assembly was only a few hundred microseconds (a few inches of motion at 1,000 fps).

The second was the "implosion method" in which a near-critical assembly of fissile material is surrounded by a layer of high explosive. When the explosive is detonated on its outer surface in such a way that a spherically converging shock wave is imposed on the fissile material compressing it by a factor between about four thirds and two, the assembly is highly supercritical by the time the shock wave reaches the center. With this method of assembly, the supercritical time may be only a few microseconds, as compared with the few hundred microseconds required in the case of the gun method. It will be obvious that with compressions of twofold or so available, some fraction of a critical mass at normal density may also be made quite supercritical by implosion. In either case—gun method or implosion—a modulated initiator, some structure containing separated layers of beryllium and polonium, for example, that is crushed by the projectile or the shock wave can be used to provide a strong source of neutrons to initiate a chain reaction when the assembly is complete.

Predetonation

In an assembly of a few tens of kilograms of highly enriched uranium (as would be required for a uranium weapon using the gun method, which provides no compression) and a neutron source in the fissile material of 2 or more neutrons/kg-s, the neutron source in the system would be on the order of 10^2 neutrons/second. With a supercritical assembly time interval of several times 10^{-4} s, the chance that a background neutron would appear during this interval would be several times 10^{-2}. The likelihood that a single neutron will initiate a chain in a mildly supercritical assembly is not very close to unity (cf. the discussion in Appendix C showing that, in a system that is nearly critical, more than 50 percent of the neutrons escape without causing a fission). It follows that the probability of predetonation of a gun-assembly of uranium can be reduced to 1 percent or so.

Plutonium-239 made in a reactor is unavoidably accompanied by some plutonium-240. The fraction plutonium-240/plutonium-239 increases with the integrated neutron flux to which the uranium-238 source material for plutonium-

239 has been exposed. The separation of plutonium from uranium is an expensive process, and it is not considered practical in obtaining plutonium in bulk quantities to extract it at an earlier stage than one at which the plutonium-240 content has reached a level of a few percent. Indeed, "weapons-grade" plutonium is defined as material having no more than 7 percent plutonium-240. The neutron source from spontaneous fission in plutonium-240 is 10^3 neutrons/gm-sec. Thus the neutron source in plutonium with only 1 percent plutonium-240 is 10 neutrons/gm-sec, and with a few kilograms of material having several percent plutonium-240, the neutron source will be somewhat larger than 10^5 neutrons/s. With such a source, and a supercritical assembly interval of several microseconds as in an implosion assembly, on the average several tenths of a neutron will appear during the interval. It requires about 10 neutrons to provide a 99 percent probability of initiating a chain in a mildly supercritical system, so the predetonation probability in such an assembly can be seen to be about a few percent.

In a gun-type system, with no compression, two or three times as much plutonium must be used as in an implosion-type. The supercritical interval is about a hundred times longer, so several hundred neutrons will appear during the supercritical interval. This is enough to assure predetonation quite early in the assembly process, and, for this reason, plutonium cannot be used effectively in the gun method.

BOOSTED WEAPONS

A booster is a fission device containing a small amount of deuterium-tritium (D-T) gas at the center. As the chain reaction proceeds, heating the fissile material, it can get to the stage at which the temperature of the fissile material, and the adjacent gas in the middle, is in the neighborhood of a kilovolt (10 million°C). At about this point, a thermonuclear reaction (deuteron plus tritium combining to yield a neutron plus an alpha particle plus 17 MeV of energy) will be initiated, which, once it is started, proceeds extremely rapidly. The energy released will be of little consequence, being overshadowed by the energy already released by fissions; but the number of neutrons produced may exceed the number otherwise present in the system. Being introduced quite independently of the progress of the chain reaction, and in a near-instantaneous pulse, the neutrons increase the rate of fissions very sharply, with the result that the yield ultimately realized may be several-fold larger than it would have been without the "boosting."

As a consequence of employing this technique, it has been possible to obtain a larger yield from a device of a given size and cost in fissile material, to obtain the same yield from a smaller amount of fissile material, and—most importantly—to obtain a desired yield from devices reduced in size and weight. Almost all weapons produced since about 1960 have been boosted.

Apart from the advantages in weapon size and weight and the direct importance of that with respect to delivery systems, the wide-scale use of boosting has had several other consequences.

• Hydrogen reacts readily with uranium or plutonium to form the solid hydride, so the D-T gas is stored at high pressure in a steel reservoir and released into the pit only at the time the weapon is to be fired. Because of the radioactive decay of tritium (a half-life of 12.3 years) these reservoirs have to be returned every few years to be recharged. Also, because of the tritium decay, it is necessary to produce new tritium to maintain the stockpile at a constant level.

• The polonium-beryllium modulated initiator previously referred to in connection with a pure fission device had a short shelf life because of the 138-day half-life of polonium-210, and the replacement of those was once a major activity at the Mound facility. In boosted weapons the initiation function has been taken over by electrically powered neutron generators obtained from the Pinellas Plant.

• Also, in the early pure fission implosion devices, the fissile material was kept outside the high explosive and only installed in place in preparation for firing. At least partly because of the geometrical complexity of the gas reservoir and transfer system, the active material in boosters is stored in place in the high explosive. This leads to the requirement for "one-point safety"—the requirement, that is, that should the high explosive be accidentally detonated at any one point (as a result, for example, of being dropped from a height, or struck by a projectile, or exposed to a fire) there must be an extremely low probability of generating any appreciable nuclear yield.

• While it would be possible to design a new pure fission device and, by a combination of non-nuclear experiments and calculation, predict the yield with very high confidence without resorting to full-scale test, this does not seem to be possible with respect to a new booster design. The booster yield depends very strongly on the state of the D-T gas at the time it may burn and on the extent of its burning; and since these conditions develop only in the course of the nuclear explosion, they are not subject to observation or confirmation by any non-nuclear experiment. For different reasons, the need for testing with actual fissile material would also apply to confirming the one-point safety of a new booster design.

THERMONUCLEAR WEAPONS

Weapons with yields much larger than 10 kilotons, or so, would probably make use of a thermonuclear design, or H-bomb. Such devices have been described as having two separate nuclear components mounted inside a case. One component, designated as the "primary," would be a fission device, most probably of the booster type because of the invariable interest in reducing the overall weight and size to the smallest feasible level consistent with the objective specified for the weapon. The other component—the "thermonuclear capsule" or "secondary"—is designed to provide almost all the total yield specified for the weapon. It consists of a mass of solid Li6D enclosed in a layer, or capsule, of some heavy metal. Since the fast (14 MeV) neutrons produced in the burning of Li6D can cause fission in uranium-238 (the most common isotope in uranium), the capsule would most probably be made of uranium, or even depleted uranium.

When the primary, with a yield of a few kilotons, say, is fired, its energy will distribute itself very rapidly throughout the volume inside the outer case and surrounding the secondary capsule. Since the case will have to fit inside some delivery vehicle, its volume is unlikely to be much larger than 1 m^3, and it may be considerably less than that. At least for a short time—until the case can be swept away—the energy density surrounding the secondary capsule will be of the order of a kiloton high explosive equivalent per cubic meter, or most probably more. The density of chemical high explosive is about 1.6 g/cm^3, so 1,000 tons of high explosive occupies a volume of about 600 m^3, and the energy provided by chemical high explosive will be about 1 kiloton per 600 m^3. The energy density, and pressure, that the explosion of the primary provides inside the case may be seen, then, to be 1,000, or so, times larger than those provided by chemical high explosives. The secondary is consequently subjected to an extremely violent implosion which will result in compressions and densities of the thermonuclear and wall materials very much larger than chemical high explosive could impose. Such conditions are favorable for a rapid thermonuclear burning of the Li6D; and the energy from this, along with that from the fissions induced in the wall, determines the yield of the weapon.

Apart from calculating the progress of processes just referred to, a main problem for the designer will be to conform to the shape and dimensional constraints imposed by the characteristics of the delivery vehicle in question, while at the same time striving to meet the conflicting desires of the military customer that the weight be reduced and the yield increased as much as possible. Subsequently, the fabricator will have to meet unusually stringent requirements on dimensional tolerances as well as on the composition, purity, and uniformity of the materials.

Appendix F
Charge to the Committee

Excerpts from the Defense Authorization Act of
Fiscal Year 1988
Public Law 100-180

Sec. 3134. INTERIM OVERSIGHT OF SAFETY OF THE NUCLEAR WEAPONS COMPLEX

(a) REQUIREMENT FOR REVIEW AND REPORT.--(1) The Secretary of Energy shall request the National Academy of Sciences to conduct two reviews on the status of the nuclear weapons complex and submit a report on each review. Each such report shall include--

(A) a consideration of safety and technical issues at current facilities and a discussion of steps that would enhance the safety of operation of those facilities;

(B) a consideration of the environmental impact of the operation of those facilities;

(C) an estimation of the approximate useful lifetime of existing reactors; and

(D) findings and recommendations.

(2) The reports shall be submitted concurrently to the Committees on Armed Services of the Senate and House of Representatives and the Secretary not later than December 1, 1988, and December 1, 1989.

Department of Energy
Washington DC 20585

February 4, 1988

Honorable Frank Press
President
National Academy of Sciences
Washington, D. C. 20418

Dear Dr. Press:

On behalf of the Department of Energy and in response to Section 3134 of the Defense Authorization Act of 1988 (P. L. 100-180), I am requesting that the Academy undertake two reviews of the Department's nuclear weapons complex. Each of the two reviews should yield reports that include the following:

- a consideration of safety and technical issues at current facilities and a discussion of steps that would enhance the safety of operation of those facilities;
- a consideration of the environmental impact of the operations of those facilities;
- an estimation of the approximate useful lifetime of existing reactors; and
- findings and recommendations.

In addition to the Secretary of Energy, the reports shall be submitted concurrently to the House and Senate Armed Services Committees. The due dates for submission of the two reports are December 1, 1988 and December 1, 1989, respectively.

I have designated Troy Wade, Acting Assistant Secretary for Defense Programs, as the Department's point of contact for this effort. I have asked Mr. Wade to make himself available to meet with you, at your earliest convenience, to discuss the details of the scope and schedule.

continued

2

By performing the two reviews, the Academy will play a key role in assisting the Department to address Congressional concerns and questions regarding oversight of the nuclear weapons complex.

We look forward to working with the Academy on this effort over the next 2 years and hope that both the Department and the Academy can find ways to expedite its start.

Yours truly,

Joseph F. Salgado
Under Secretary

cc:
Honorable Sam Nunn
Chairman, Committee on Armed Services
United States Senate

Honorable Les Aspin
Chairman, Committee on Armed Services
House of Representatives

Department of Energy
Washington, DC 20585

MAY 13 1988

Honorable Sam Nunn
Chairman, Committee on Armed Services
United States Senate
Washington, DC 20510

Dear Mr. Chairman:

On February 4, 1988, the Department requested the National Academy of Sciences to conduct two reviews of the nuclear weapons complex. This request was made pursuant to Division C, Title I, Section 3134 of the Defense Authorization Act of 1988 (P.L. 100-180). In our request to the Academy, we asked for two reports that addressed the following:

- a consideration of safety and technical issues at current facilities and a discussion of steps that would enhance the safety of operation of those facilities;
- a consideration of the environmental impact of the operations of those facilities;
- an estimation of the approximate useful lifetime of existing reactors; and
- findings and recommendations.

Along with the submission of the reports to the Secretary, we requested concurrent submission to the House and Senate Appropriations and Armed Services Committees. The due dates requested for the two reports were December 1, 1988, and December 1, 1989, respectively, as outlined in Section 3134.

The Academy has responded with a proposal that varies from our request in two ways. First, the Academy's review will not cover the defense production reactors; and secondly, the Academy will report only on December 1, 1989. The Academy's proposed departures are designed to avoid an unnecessary duplication of effort with the Department's Advisory Committee on Nuclear Facilities Safety, and provide the time necessary to conduct a thorough and complete review as intended by Section 3134.

continued

2

We believe the Academy's approach will meet the intent of Section 3134 and at the same time compliment the Department's initiatives on nuclear facilities safety. If you require additional information or feel further discussion is required, please contact me at your earliest convenience.

Sincerely,

Troy E. Wade II
Acting Assistant Secretary
for Defense Programs

cc:
Honorable John W. Warner
Ranking Minority Member
Committee on Armed Services
United States Senate
Washington, DC 20510

NATIONAL RESEARCH COUNCIL

COMMISSION ON PHYSICAL SCIENCES, MATHEMATICS, AND RESOURCES
COMMISSION ON ENGINEERING AND TECHNICAL SYSTEMS

2101 Constitution Avenue Washington D.C. 20418

COMMITTEE TO PROVIDE INTERIM OVERSIGHT OF
THE DOE NUCLEAR WEAPONS COMPLEX

November 30, 1988

Honorable Sam Nunn
Chairman, Committee on Armed Services
United States Senate
Washington, DC 20510

Honorable Les Aspin
Chairman, Committee on Armed Services
House of Representatives
Washington, DC 20515

Honorable John S. Herrington
Secretary, Department of Energy
Forrestal Building
Washington, DC 20585

Gentlemen:

I am pleased to report on the activities and current status of the National Research Council Committee to Provide Interim Oversight of the DOE Nuclear Weapons Complex.

Following hearings sponsored by the Senate Governmental Affairs Committee and the Senate and House Armed Services Committees, and in response to Division C, Title 1, Section 3134 of the Defense Authorization Act of 1988, the Department of Energy requested the National Academy of Sciences to undertake a review of the department's nuclear weapons complex. That request is contained in a February 4, 1988 letter from Under Secretary Joseph Salgado to President Frank Press of the National Academy of Sciences. The letter is attached as Appendix 1.

As a result of discussions with DOE, the actual charge to the Committee differs somewhat from that proposed in the February letter. The final charge was modified primarily to avoid unnecessary duplication of effort with DOE's Advisory Committee on Nuclear Facility Safety. Dr. Troy Wade, DOE's Acting Assistant Secretary for Defense Programs, apprised you of this in a May 13, 1988 letter, which is attached as Appendix 2.

continued

Page 2.
November 30, 1988

The committee is conducting an 18-month study to lay the groundwork for operation of the legislatively mandated, permanent independent board that will provide oversight of DOE's nuclear weapons complex. The committee has been asked to examine safety and environmental issues at a variety of facilities, but will not consider the defense production reactors; safe handling of nuclear weapons; the waste isolation pilot plant; transportation safety; and activities at the Nevada Test Site. The final report, scheduled for December 1, 1989, will include:

--A consideration of safety and technical issues at current facilities and a discussion of steps that would enhance the safety of operation of those facilities;

--A consideration of the environmental impact of the operations of those facilities;

--Findings and recommendations.

A committee of 22 experts has been appointed by the Academy. The members provide a balance of expertise in chemical processing, criticality safety, environmental assessment, explosives safety, fire safety, laboratory management, materials handling, materials science, nuclear safety, pulsed power safety, remote systems technology, and seismic risk. A list of the members, including concise biographical sketches, is attached as Appendix 3.

To date the committee has held two meetings. The first meeting was held at the National Academy of Sciences in Washington, D.C. on August 22-23 and the second was at the Hanford site in Richland, Washington on October 24-26. At its first meeting, the committee received briefings from Department of Energy headquarters officials; from the Chairman of DOE's Advisory Committee on Nuclear Facility Safety; from staff of the U.S. Environmental Protection Agency; and from staff of the U.S. General Accounting Office.

At its meeting in Hanford in October, staff of DOE's Richland Operations Office provided the committee with an introduction and overview of operations in the Hanford 200 areas, and personnel of the Westinghouse Hanford Company briefed us on various aspects of management and operations. The Committee toured the PUREX Plant, the Plutonium Finishing Plant, the 200-Area Tank Farms and B Plant. The Committee also met separately with staff of the Washington State Department of Ecology.

continued

Page 3.
November 30, 1988

In addition to these activities, committee staff have made several subsequent visits to Hanford to review available documents and interview staff of DOE, Westinghouse, Pacific Northwest Laboratories, and the Washington State Department of Ecology.

Plans for additional meetings and site visits are currently under consideration.

The committee is aware of the magnitude of its task and the importance of these facilities to the future health, safety, and security of the nation. Mindful of this, we will make a determined effort to provide a thorough and timely report by December 1, 1989, that will be useful to the Executive and Legislative branch policy makers of the government, and to the permanent oversight board.

Sincerely yours,

Richard A. Meserve
Chairman

cc. Honorable John Glenn
Mr. Joseph Salgado
Dr. Troy Wade

NATIONAL RESEARCH COUNCIL

COMMISSION ON PHYSICAL SCIENCES, MATHEMATICS, AND RESOURCES
COMMISSION ON ENGINEERING AND TECHNICAL SYSTEMS

2101 Constitution Avenue Washington D.C. 20418

COMMITTEE TO PROVIDE INTERIM OVERSIGHT OF
THE DOE NUCLEAR WEAPONS COMPLEX

February 27, 1989

The Honorable Sam Nunn
Chairman, Committee on Armed Services
United States Senate
Washington, D.C. 20510

The Honorable Les Aspin
Chairman, Committee on Armed Services
United States House of Representatives
Washington, D.C. 20515

Admiral James D. Watkins
Secretary-Designate
Department of Energy
Forrestal Building
Washington, D.C. 20585

Gentlemen:

This is to follow up on my letter of November 30, 1988, reporting on the activities and status of the National Research Council's Committee to Provide Interim Oversight of the DoE Nuclear Weapons Complex. Since then, on January 26, 1989, I testified before the Senate Governmental Affairs Committee on behalf of the National Academy of Sciences. Copies of both my November letter and my January testimony are enclosed.

The Committee's visits in October to the Hanford Reservation in Washington; in January to the Y-12 Facility in Oak Ridge, Tennessee; and in February to the Idaho Chemical Processing Plant will be complemented by an aggressive schedule of visits over the coming months: a March visit to the Rocky Flats Plant in Colorado; April visits to the Lawrence Livermore National Laboratory in California, the Los Alamos National Laboratory and the Sandia National Laboratory in New Mexico, and the Pantex Plant in Texas; and a May visit to the Savannah River Plant in South Carolina. This schedule is driven by the large scope of the assignment -- the examination of safety and environmental issues at the defense weapons complex -- and by the deadline for our final report of December 1989.

The National Research Council is the principal operating agency of the National Academy of Sciences and the National Academy of Engineering to serve government and other organizations

continued

The Honorable Sam Nunn
The Honorable Les Aspin
Admiral James D. Watkins
February 27, 1989
Page 2

Although the Committee will visit only nine of the fourteen major facilities in the nuclear weapons complex that are encompassed by our charge, we are seeking, within the constraints of available time and resources, to examine a representative cross section of the complex. We believe our examination of a subset of the facilities will provide an adequate foundation for our report.

Sincerely yours,

Richard A. Meserve
Chairman

Enclosures

cc: The Honorable John Glenn
Dr. Frank Press

References

American Nuclear Society/American National Standards Institute. 1975-1985. Nuclear facility safety standards 8.1-8.12. La Grange Park, Ill.

Brown, C.L. 1987. Forecast of criticality experiments needed to support U.S. DOE contractor operations: 1987–1992. NCT-03, Nuclear Criticality Information System, Lawrence Livermore National Laboratory. Livermore, Calif.

Cleveland, J.M. 1979. The Chemistry of Plutonium. La Grange Park, Ill.: American Nuclear Society.

Comar, C.L., et al. 1976. Plutonium: Facts and Inferences. Report no. EPRI EA-43-SR. Electric Power Research Institute, Palo Alto, Calif. (Available from National Technical Information Service, Springfield, Va., as PB-260 584.)

Cook-Mozaffari, P., F.L. Ashwood, T. Vincent, et al. 1987. Cancer incidence and mortality in the vicinity of nuclear installations. England and Wales, 1950–1980. Studies on Medical and Population Subjects, No. 51. London: Her Majesty's Stationery Office.

Darby, S.C., and R. Doll. 1987. Fallout, radiation doses near Dounreay, and childhood leukaemia. British Medical Journal 294:603-607.

Department of Energy. 1981. A Report on a Safety Assessment of Department of Energy Nuclear Reactors. DOE/US-0005. Report of the Crawford Committee. Washington, D.C.

Department of Energy. 1985. Department of Energy National Environmental Research Parks. DOE/ER-0246. Office of Energy Research. Washington, D.C.

Department of Energy. 1987. Determination to establish Advisory Committee on Nuclear Facility Safety. Federal Register 52(222):44208.

Department of Energy. 1988. Environment, safety and health report for the Department of Energy Defense Complex. Internal document prepared July 1, 1988. Washington, D.C.

Department of Energy. 1989a. Environmental Restoration and Waste Management: Five-Year Plan. Washington, D.C.

Department of Energy. 1989b. Secretary of Energy Notice: Departmental Organization and Management Arrangements, 5/19/89. Washington, D.C.

Department of Energy. 1989c. Secretary of Energy Notice: Departmental Organization and Management Arrangements, 9/28/89. Washington, D.C.

Department of Energy. 1989d. Predecisional Draft II: Environmental Restoration and Waste Management Five-Year Plan, July 3, 1989. Washington, D.C.

Forman, D., P. Cook-Mozaffari, S. Darby, G. Davey, I. Stratton, R. Doll, and M. Pike. 1987. Cancer near nuclear installations. Nature 329:499-505.

Jablon, S., B.J. Stone, Z. Hrubec, and J.D. Boice, Jr. 1988. Cancer Mortality in the Environs of Nuclear Facilities. National Cancer Institute. U.S. Department of Health and Human Services, Washington, D.C. Photocopy.

Kinlen, L. 1988. Evidence for an infective cause of childhood leukaemia: Comparison of a Scottish new town with nuclear reprocessing sites in Britain. The Lancet ii:1323-1327.

National Academy of Public Administration. 1986. Summary Report of the NASA Management Study Group: Recommendations to the Administrator, National Aeronautics and Space Administration. Washington, D.C.

National Academy of Sciences. 1985. Pp. 229-230 in Nuclear Arms Control: Background and Issues. Washington, D.C.: National Academy Press.

National Council on Radiation Protection and Measurements. 1987. Ionizing Radiation Exposure of the Population of the United States. NCRP Report No. 93. Bethesda, Md.

National Research Council. 1987. Safety Issues at the Defense Production Reactors. Committee to Assess Safety and Technical Issues at DOE Reactors. Washington, D.C.: National Academy Press.

National Research Council. 1988a. Role of the Primary Care Physician in Occupational and Environmental Health. Institute of Medicine. Washington, D.C.: National Academy Press.

National Research Council. 1988b. Safety Issues at the DOE Test and Research Reactors. Committee to Assess Safety and Technical Issues at DOE Reactors. Washington, D.C.: National Academy Press.

National Research Council. 1988c. Health Effects of Radon and Other Internally Deposited Alpha-Emitters: BEIR IV. Board on Radiation Effects Research. Washington, D.C.: National Academy Press.

National Research Council. 1989a. Review comments on predecisional Draft II of DOE's Environmental Restoration and Waste Management Five-Year Plan. Letter report, 8/3/89. Board on Radioactive Waste Management. Washington D.C.

National Research Council. 1989b. Summary Report 1987: Doctorate Recipients from United States Universities. Office of Scientific and Engineering Personnel. Washington, D.C.: National Academy Press.

National Research Council. 1989c. Drinking Water and Health. Vol. 9, Selected Issues in Risk Assessment. Board on Environmental Science and Technology. Washington, D.C.: National Academy Press.

National Research Council. 1989d. Improving Risk Communication. Committee on Risk Perception and Communication. Washington, D.C.: National Academy Press.

National Research Council. 1990. The Effects on Population of Exposure to Low Levels of Ionizing Radiation: BEIR V. Board on Radiation Effects Research. Washington, D.C.: National Academy Press.

National Safety Council. 1987. Work Injury and Illness Rates. Chicago, Ill.

ORAU. 1988. Department of Energy nuclear workers study. Oak Ridge Associated Universities, Oak Ridge, Tenn. October 1. Photocopy.

Pacific Northwest Laboratory. 1988. Environmental Monitoring at Hanford for 1987. PNL-6464. Richland, Wash.

Pacific Northwest Laboratory. 1989. Radiation Exposures for DOE and DOE Contractor Employees, 1987. Twentieth annual report prepared for the U.S. Department of Energy Assistant Secretary for Environment, Safety, and Health. Richland, Washington.

Paxton, H.C., and N.L. Pruvost. 1986. Critical Dimensions of Systems Containing ^{235}U, ^{239}Pu, and ^{233}U. (Revised.) LA-10860-MS, Los Alamos National Laboratory, N. Mex.

Public Health Service. 1984. Vital Statistics of the United States 1980. National Center for Health Statistics, Hyattsville, Md.

Rogers, W.P. 1986. Report to the President by the Presidential Commission on the Space Shuttle Challenger Accident. Washington, D.C.: U.S. Government Printing Office.

Scientech Inc. 1989a. Technical Review of Plutonium Holdup in the Plutonium Finishing Plant Ventilation System. Prepared for the Department of Energy.

Scientech Inc. 1989b. An Assessment of Critical Safety at the Department of Energy, Rocky Flats Plant, Golden, Colorado, July–September 1989. Prepared for the Department of Energy.

SRS. 1989. Draft SRS Environmental Information Booklet for the National Academy of Sciences Site Review. Savannah River Site, Aiken, S.C.

Till, J.E. In press. Reconstructing Historical Exposure to the Public from Environment Sources. In Proceedings of the 25th Annual Meeting of the National Council on Radiation and Measurements. Bethesda, Md.

Tuck, J.C. 1989. Statement before the panel, Subcommittee on Procurement and Military Nuclear Systems, Armed Services Committee, U.S. House of Representatives. 7/18/89. Washington, D.C.

Wachholz, B.W. In press. Overview of the National Cancer Institute's activities related to exposure of the public to fallout from the Nevada test site. Health Physics.

Abbreviations

ACNFS	Advisory Committee on Nuclear Facility Safety
AEC	Atomic Energy Commission
ALARA	as low as reasonably achievable
ANS	American Nuclear Society
ANSI	American National Standards Institute
ASDP	Assistant Secretary for Defense Programs
ASEH	Assistant Secretary for Environment, Safety, and Health
ASNE	Assistant Secretary for Nuclear Energy
BEIR	biological effects of ionizing radiations
BEST	Board on Environmental Science and Technology
BRWM	Board on Radioactive Waste Management
CAM	continuous air monitor
CDC	Centers for Disease Control
CEDR	Comprehensive Epidemiologic Data Repository
CERCLA	Comprehensive Environmental Response, Compensation, and Liability Act
CRERP	Committee on the DOE Radiation Epidemiologic Research Programs
CWA	Clean Water Act
DNFSB	Defense Nuclear Facilities Safety Board
DOD	Department of Defense
DOE	Department of Energy
DOR	direct oxide reduction
DP	Office of Defense Programs

EH	Office of Environment, Safety, and Health
EML	Environmental Measurements Laboratory (New York City)
EPA	Environmental Protection Agency
ERDA	Energy Research and Development Administration
FMPC	Feed Materials Production Center
HEPA	high-efficiency particulate air filter
IARC	International Agency for Research on Cancer
ICPP	Idaho Chemical Processing Plant
ICRP	International Council on Radiation Protection
INEL	Idaho National Engineering Laboratory
LACAF	Los Alamos Critical Assembly Facility
LANL	Los Alamos National Laboratory
LEAF	Legal Environmental Assistance Foundation
LLNL	Lawrence Livermore National Laboratory
NAPA	National Academy of Public Administration
NASA	National Aeronautics and Space Administration
NFPA	National Fire Protection Association
NPR	new production reactors
NPS	National Priority System
NRC	National Research Council (Nuclear Regulatory Commission always spelled out)
NSR	new special recovery
NWC	Nuclear Weapons Complex
OHER	Office of Health and Environmental Research
OSEP	Office of Scientific and Engineering Personnel
OSHA	Occupational Safety and Health Administration
OSR	operational safety requirement
PFP	Plutonium Finishing Plant
RCRA	Resource Conservation and Recovery Act
RESL	Radiological and Environmental Sciences Laboratory (Idaho Falls)
RFP	Rocky Flats Plant
SEN	Secretary of Energy Notice
SIS	special isotope separation
SNL	Sandia National Laboratory
SPEERA	Secretarial Panel for Evaluation of Epidemiologic Research Activities
SREL	Savannah River Ecology Laboratory
SRS	Savannah River Site
WINCO	Westinghouse Idaho Nuclear Company